SEX, SKULLS, AND CITIZENS

SEX, SKULLS, AND CITIZENS

Gender and Racial Science in Argentina (1860–1910)

ASHLEY ELIZABETH KERR

Vanderbilt University Press
Nashville, Tennessee

Nashville, Tennessee 37235

First printing 2020

Library of Congress Cataloging-in-Publication Data

Names: Kerr, Ashley Elizabeth, 1984- author.
Title: Sex, skulls, and citizens : gender and racial science in Argentina (1860–1910) / Ashley Elizabeth Kerr.
Description: Nashville : Vanderbilt University Press, [2019] | Includes bibliographical references and index. | Summary: "Based on analysis of a wide variety of late-nineteenth-century sources, this book argues that indigenous and white women shaped Argentine scientific racism as well as its application to projects aiming to create a white, civilized nation. The writers studied here, scientists, anthropologists, and novelists, including Estanislao Zeballos, Lucio and Eduarda Mansilla, Ramon Lista, and Florence Dixie, reflect on indigenous sexual practices, analyze the advisability and effects of interracial sex, and use the language of desire to narrate encounters with indigenous peoples as they try to scientifically pinpoint Argentina's racial identity and future potential"-- Provided by publisher.
Identifiers: LCCN 2019027141 (print) | LCCN 2019027142 (ebook) | ISBN 9780826522719 (hardback) | ISBN 9780826522726 (paperback ; alk. paper) | ISBN 9780826522733 (ebook)
Subjects: LCSH: Ethnology—Argentina—History—19th century. | Indigenous peoples—Argentina—History—19th century. | Race discrimination—Argentina—History—19th century. | Sex discrimination—Argentina—History—19th century. | Sex—Argentina—History—19th century. | White nationalism—Argentina—History—19th century. | Argentina—Race relations—History—19th century.
Classification: LCC GN564.A7 K47 2019 (print) | LCC GN564.A7 (ebook) | DDC 305.800982—dc23
LC record available at https://lccn.loc.gov/2019027141
LC ebook record available at https://lccn.loc.gov/2019027142

This book is dedicated to Jason and Emily. Thank you so much for your constant support, love, and timely distractions.

CONTENTS

ACKNOWLEDGMENTS

Funding for this project came from the University of Idaho, including a Seed Grant and a CLASS summer travel grant. I am also grateful to the Prindle Institute for Ethics and their summer seminar, which provided a place to write and discuss my ideas.

I could not have completed this book without the help of Máximo Farro at the Museo de La Plata, who gave me access to the archives and directed me towards numerous sources. In Buenos Aires, the staff at the Archivo General de la Nación, Biblioteca del Congreso de la Nación, and Biblioteca Nacional were helpful even when I had no idea what I was looking for. Many thanks as well to the Fundación del Museo de La Plata for permission to publish images from their archives.

Numerous friends, family members, and colleagues made this book a reality by reading chapters, talking out ideas, and providing support. Thank you to my parents and sister, to Rachel Halverson and my other colleagues in MLC, and to Rebecca Scofield, Sean Quinlan, Matthew Fox-Amato, Tara MacDonald, Erin James, Stefanie Ramirez, Kara Yedinak, James Riser, Allison Libbey, Natalie McManus-Chu, Katherine Karr-Cornejo, Stephen Silverstein, Adriana Rojas, Miguel Fernández, Fernando Operé, Mané Lagos, David Gies, Anna Brickhouse, Gustavo Pellón, Mariela Eva Rodriguez, the participants at the 2018 CHAA conference in Buenos Aires, Jennifer Watson Wester, and many others. Extra-special thank yous for Ruth Hill, who has been my one of my biggest cheerleaders for over a decade, and Janice North, who read the entire manuscript and gave me the feedback that made it what it is today.

Thank you as well to Zack Gresham, the Vanderbilt University Press team, and the anonymous peer reviewers whose comments were so valuable.

INTRODUCTION

Scientific Engagements

Women, Sex, and Racial Science

In the midst of hundreds of pages of ethnographic observations of the Tehuelche and Manzanero Indians in his 1879 account of exploring southern Patagonia, Francisco P. Moreno offhandedly remarks that he had been asked to provide medical aid to the *cacique* (chieftain) Shaihueque's niece, "Chacayal's daughter, who was at the same time my betrothed."[1] At approximately the same time, the Tehuelche chieftain Pecho Alegre extended an open invitation to marry his daughter to the explorer Ramón Lista: "Whenever you want, say; I give girl *free*."[2] Four years later, Estanislao Zeballos also briefly alluded to a similar encounter, noting that he left a Ranquel camp after hours of ethnographic observation "smelling like colt and refusing to return in order to accept the wife that they offered me and who was the Indian woman *Epuloncó*." It is clear that none of the men is interested in such a union. Moreno's tone is sarcastic, Lista refuses, and after directly registering his denial, Zeballos writes, "I ran to change all my clothes and wash my hair. One does not cultivate the society of the *toldos* [indigenous encampment] with impunity!"[3] Instead, the anthropologists' reports of the tribes' offers of wives and their own refusals contribute to their narration of white Argentine racial superiority vis-à-vis the cultures they study by insisting on their ultimate control over the situation. Although depicted as transactions between men representative of their respective races, women's bodies are essential to these encounters.

After winning independence from Spain, the new nations in Latin America worked to solidify control over their territories, identify their populaces, and formulate productive national identities. As John Chasteen argues, in most of Latin America, the nation emerged decades after the creation of independent states and only as the result of intentional and often arduous efforts.[4] Race was key to these projects: in Latin America, "national identities have been constructed in racial terms" while "definitions of race have been shaped by processes of nation building."[5] In the case of nineteenth-century Argentina, the racial elements in question were primarily Creoles of European descent, rich and varied indigenous cultures, and the mixed-race mestizos that resulted from their unions.[6]

By the late nineteenth century, anthropology, ethnography, comparative anatomy, and other forms of racial science were the most authoritative methodologies for physically and symbolically shaping the nation. Scholars have generally understood the transformation via science of Argentina's inhabitants into racialized citizens or outsiders as an ungendered process. Most agree that white men went to the frontier to study indigenous men and found them barbarous, and those findings allowed white male politicians and military men to realize actions designed to exterminate or assimilate the *indios de lanza* (male warriors). Although the participants are identified as male, both the construction and the consequences of this masculinity are elided in most scholarly narratives. Women exist marginally, if at all. Furthering this erasure, many of the key anthropological texts of the period focus on Patagonia, a region historically associated with masculinity.[7]

Nonetheless, the episodes Moreno, Lista, and Zeballos recount demonstrate that nineteenth-century Argentine scientific projects were not just racial encounters, but also gendered ones. Analyzing a wide variety of late nineteenth-century sources, this book argues that indigenous and white women shaped Argentine scientific racism as well as its application to projects aiming to create a white, civilized nation. Often, although not always, these encounters were conditioned by sexual intercourse as practice and discourse. As the chapters show, "women" refers to real-life figures who interacted with, were studied by, or challenged the male explorer-scientists dedicated to constructing knowledge regarding race in the Argentine context. It also refers to the numerous imaginary or symbolic female figures used to explore anxieties about race and national identity. As sexual partners, skulls to study, and potential citizens in formation, women were at the heart of the nineteenth-century scientific enterprise.

Moreno, Lista, and Zeballos's fleeting engagements hint at some of the patterns that will emerge over the course of this book. First, contrary to present-day faith in the impersonal and objective nature of science, sex is a constant presence in the late nineteenth-century texts. Zeballos's narrative juxtaposition of his rejection of Epuloncó, the smell of horse, and the insistence that he scrubbed himself clean immediately after leaving make clear that implicit behind the offer and rejection of indigenous wives is the suggestion of sex in all its physical, messy reality. The writers I study reflect on indigenous sexual practices, analyze the advisability and effects of interracial sex, and use the language of desire to narrate encounters with indigenous peoples as they try to scientifically pinpoint Argentina's racial identity and future potential. While Moreno and Zeballos quickly shut down any possibility of their own participation in interracial sexual encounters, others engaged in more extensive narrative flirtations. In the passage quoted above, Ramón Lista refused Pecho Alegre's offer with a polite "Not now, compadre," foreshadowing his sexual relationship with a Tehuelche woman named Clorinda Coile a decade later.[8]

As the preponderance of musings on sex suggests, women also played a larger role in the production, diffusion, and deployment of scientific ideas about race then previously supposed. Sometimes we can recover their names, as in the case of Zeballos's offered wife, Epuloncó; others are remembered only in relation to men (Moreno's "daughter of"), or through the generic label of woman or girl. Indigenous women's bodies were photographed and measured in order to create knowledge about race while both their bodies and those of white women were sites for focalizing concerns about interracial desire and the physical reproduction of national citizens. Indigenous people also used indigenous women's bodies to diplomatically navigate encounters with the creole world. The women offered to Zeballos, Lista, and Moreno were not just signs of indigenous gratitude or awe for civilization, for across the Americas "Native women were routinely offered to and accepted by outsiders in order to initiate and maintain strategic connection."[9] Indeed, in 1871 British explorer George Chaworth Musters explicitly noted the tit-for-tat of "matrimonial entanglements" in Argentine Patagonia, insisting that his potential wife's community had encouraged her to approach him in order to gain access to his firearms and support.[10]

Women also took on more active roles. Although we frequently view the destruction of Argentine indigenous communities as happening in masculine military and political spaces, this book demonstrates that some white women helped separate indigenous families and were charged with the gendered re-education in creole domesticity of many indigenous women and children. Others, including the Argentine Eduarda Mansilla and the Scotswoman Florence Dixie, offered opinions on masculine scientific-political endeavors in texts whose intended audiences included women and girls.

Indigenous people, including women, also surpassed the stereotype of passive object and actively contributed to the development of Argentine racial science. Some acted as ethnographic informants and shaped Argentine representations of their tribes. Occasionally these participations are explicitly acknowledged; others can only be assumed. While the nineteenth-century explorers appear to take their information seriously and at face value, there is evidence in the texts that indigenous women were instrumental in resisting Argentine scientific and colonial projects. This book collects cases in which female indigenous informants and mediators disrupted scientific representations, often by exploiting sexual desire and stereotype. By interrogating their participation, we can ask where indigenous people, including women, might have had agency when complete passivity was previously assumed.

The theories of nineteenth-century Argentine scientific racism had real effects on the people living in the territory. The ties between science and military-political actions against indigenous peoples were strong and bidirectional. The national and provincial governments sponsored scientific expeditions, including the ones Moreno, Lista, and Zeballos were undertaking when offered wives. Zeballos's 1878 *La conquista de quince mil leguas* was commissioned by the national government as a blueprint for General Roca's Conquest of the Desert, which resulted in the death or imprisonment of over fourteen thousand Argentine Indians in the period from April to July 1879.[11] Similarly, the anthropologists relied on military campaigns to help collect indigenous bones from the furthest corners of the territory.[12] Zeballos, Moreno, Lista, and their peers also served as elected representatives, governors, and/or official experts. As this book will show, these dual roles as scientists and politicians allowed anthropological knowledge to rapidly influence concrete action, including a number of policies aimed specifically at disciplining gendered indigenous bodies.

In addition to these ties to policy, scientific debates often spilled out into the literary world. Zeballos's experiences on the frontier, including the encounter with Epuloncó, inspired him to write a trilogy of works that became progressively more imaginative. He was not alone: with the exception of Moreno, all of the anthropologists I study wrote fictional texts about Argentine indigenous peoples during the period in which they were carrying out their scientific activities. Running the gamut from novels to epic poetry, these texts revolve around romantic relationships that cross racial lines. While they have generally been dismissed by scholars as nonscientific and/or of poor literary quality, the sex and romance in these fictional texts highlight the racial bases of fiction, the cultural bases of racial science, and the pathways through which scientific production was popularized and integrated into the national imagination. Fictional spaces, though different from scientific texts, were equally serious sites for grappling with racial science and concerns about national identity.

Finally, while new lenses can bring pictures into focus, they can also sow confusion: the resulting image is both richer and more resistant to being tied up nicely with a bow, introducing untidiness and contradictions into the narrative. Indeed, throughout this book it is apparent that Argentine scientific racism between 1860 and 1910 was not an inflexible nor hegemonic ideology. Individuals' opinions changed over time, as well as in relation to geography and other circumstances. As in the present, people were complex and contradictory. Some white men lusted after indigenous women while declaring the need to eliminate their culture; others argued against policies inspired by their research on race because of gendered ideologies. Indigenous peoples suffered at the hands of creole society, but many also found small ways to resist even under oppressive circumstances. Although it is tempting to view the scientific racism of the late nineteenth century as a story of heroes, villains, and victims, regendering the figures involved makes abundantly clear that such a facile division is impossible.[13] Through this framework, we can investigate complexity without negating the racism and incredible human tragedy of the period.

Creole-Indigenous Interactions, Science, and Identity

In neighboring Paraguay, Guaraní is an official language spoken by up to 90 percent of the population. In Chile, the Mapuche still play an important role in both the national imaginary and present-day politics. In contrast, Buenos Aires is often considered the Paris of the South, and Argentines call themselves "the children of the boats" ("los hijos de los barcos"), alluding to the importance of European immigration to national identity. Nonetheless, the territory on both sides of the Río de la Plata was home to rich indigenous cultures prior to the arrival of the Spanish. While those cultures suffered centuries of efforts to physically and culturally diminish them, today, Argentina has a larger relative and absolute indigenous population than several other Latin American nations, including Brazil.[14] This seeming paradox prompts the question of how the country arrived at this point.

Europeans arrived in the River Plate in 1516, sparking the parallel processes of violent clashes and peaceful mixture that would define race relations in the region. As theorized by Mary Louise Pratt, frontiers are not dividing lines, but instead contact zones, "social spaces where disparate cultures meet, clash, and grapple with each other."[15] In what is now Argentina, there were periods of serious *malones* (indigenous raids) on creole settlements that led to death, the loss of property, and the capture of women and children. During 1875 alone, four thousand Indians crossed the frontier and stole three hundred thousand heads of cattle, killed five hundred *cristianos* (Christians), and took three hundred captives.[16] The Argentine army reacted in kind to these raids, attacking the toldos to rescue stolen property and women and enact revenge.[17] At the same time, biological and cultural mixture were a fact of life even during the worst periods of warfare. Many indigenous groups maintained strong peaceful ties to creole society.[18] Pampas Indians supported the Argentine army in fighting British invasions in 1806 and 1807, frontier commerce was essential to both populations, and mixed-race peoples were a large and growing group.[19] A failed constitution project in 1819 attempted to establish that indigenous men in all the provinces were "perfectly free men with equal rights as all the citizens that populated them."[20] The first official Constitution in 1853 made this idea law, codifying the tense relationship between indigenous peoples as Argentines and indigenous peoples as inferior, internal enemies.

For several years after the establishment of this Constitution, Argentine politics were consumed by civil war at home and the War of the Triple Alliance against Paraguay. The 1862 election of Bartolomé Mitre as the first president of the reunified Republic and the end of the war in 1870 began to free up the resources of the military and treasury, motivating new interest in resolving the *cuestión de indios* (Indian question) once and for all. The Generation of 1837 had advocated for constructing the new nation on the basis of European (particularly French) liberal thought, immigration, urbanism, and progress. As president from 1868 to 1874, Domingo F. Sarmiento implemented policies designed to reach those goals, including open borders and free public education. The ideological descendants of Sarmiento and his allies continued this work. Often titled the Generation of '80, this group of aristocratic men—including Miguel Cané, Lucio V. Mansilla, Estanislao Zeballos, Julio A. Roca, and Eduardo Wilde—governed from 1880 to 1916, although their intellectual power began at least a decade earlier. Steeped in positivism influenced by Spencerism, they believed firmly in order and progress and sought to transform the River Plate in accordance with these principles.[21]

Many saw indigenous peoples as a stumbling block to their goals because their continued sovereignty limited Argentine territorial and economic expansion, a concern whose roots can be traced back to the colonial period. Indigenous communities also challenged intellectual and political elites' attempts to elaborate a productive national identity that could unify the nation and positively represent it to the world. The presence of so-called savages within national borders raised questions about the modernity and level of civilization the country could achieve. North American, European, and even some Latin American scholars looked at the chaos that characterized many of the new republics in the Americas and attempted to explain it as a function of an inherent racial character.[22] The Creole's racial inferiority was attributed to distinct causes, including their Spanish inheritance, the native peoples' racial qualities, and the process of miscegenation. In *Conflicto y armonías de las razas en América*, Sarmiento lamented the effects of these attitudes on the national psyche: "Are we European?—So many copper-colored faces refute us! Are we Indians?—Disdainful smiles from our fair ladies give the only response. Mixed? No one wants to be it, and there are thousands that will not want to be called neither American nor Argentine."[23] Consequently, many Argentines dedicated themselves to distancing the nation from the stain of indigeneity or otherwise defending the Creole from "the unfounded or false assessments" that led Argentines themselves to "denounce our own race and the civilization that gave us existence, blaming them exclusively as the cause of the evils that come from very different and varied circumstances."[24]

The need to resolve the perceived economic and symbolic problems posed by indigenous peoples was inextricably linked to the development of racial science in the region. Efforts to develop national science began shortly after independence and intensified in the 1860s and 1870s with the arrival of foreign scientists, the creation of new university degrees, and the founding of specialized journals.[25] Knowledge about the emerging fields of anthropology, ethnography, ethnology, and comparative anatomy began to flow into the country, even as local scholars developed their own new ways of understanding indigenous peoples and their cultures.

Lucio V. Mansilla's *Una excursión a los indios ranqueles* (1870) is often noted as one of the first ethnographies in the region, marking the beginning of a period of intense activity. In the next forty years, men such as Ramón Lista, Francisco P. Moreno, Estanislao Zeballos, Eduardo L. Holmberg, Florentino Ameghino, and Clemente Onelli explored the Argentine interior, observing native peoples and collecting bones and relics.[26] They corresponded with foreign anthropologists such as Paul Topinard and Paul Broca, subscribed to and published in US and European journals, and created institutions for the continued development of ethnography, anthropology, comparative anatomy, and other disciplines connected to race and racial development.

While classic histories of River Plate science such as those by José Babini and Miguel de Asúa have a nationalistic bent, recent scholarship has pinpointed how these anthropological activities undergirded the efforts to minimize the physical and symbolic importance of indigenous peoples in Argentine life. Unlike European scientists for whom analyzing the racial characteristics of savages was largely an intellectual question, River Plate scientists had to grapple with the fact that the savages they observed and measured were also citizens.[27] Indigenous people shared the same spaces, traded extensively with creole communities, and had long influenced the nation's gene pool through sexual relations that revealed the illusion of a definitive frontier between white and indigenous societies. As such, the theories of racial science provided a new language in which to discuss and justify centuries-old preoccupations with the economic and territorial threats posed by indigenous peoples.

At nearly exactly the same time as Argentine anthropology came into being, the Argentine government intensified its efforts to control native populations, notwithstanding their legal inclusion in the nation. On August 14, 1878, Congress approved Minister of War Julio A. Roca's plan to use military force to subjugate indigenous communities, earmarking 1,600,000 pesos for the efforts. Although this "Conquest of the Desert" had its critics, including those who thought it was too costly, impossible, or illegal, given the indigenous peoples' status as citizens, military excursions began almost immediately.[28] Roca and his men's efforts to crush tribal organizational structures, displace tribes, and force assimilation effectively destroyed the majority of indigenous communities in the Pampas and Northern Patagonia, even while indigenous individuals continued (and continue) to live in the region.[29] The official end of the campaign came in early 1885: on February 20, General Lorenzo Vintter sent a telegram to General Domingo Viejobueno declaring, "I can tell your honor that today there is no tribe in the countryside that has not been reduced voluntarily or by force."[30]

As Pablo Azar, Gabriela Nacach, Pedro Navarro Floria, and Mónica Quijada have shown, River Plate intellectuals justified violently excluding indigenous citizens from territories and national identity by manipulating the concepts of savagery, barbarism, and civilization to depict them as prehistoric and thus incompatible with progress.[31] Theories of pre-Colombian Aryanism cast contemporary indigenous Argentines as degenerate or illegitimate usurpers while implicitly arguing that non-indigenous Argentines were equal to (or greater than) their peers in Europe.[32] Underlying many of these assertions were unique syntheses of evolutionary thought. Drawing from Darwin, Spencer, Lamarck, Haeckel, and others, these theories further naturalized the "extinction" of indigenous peoples and provided hope that the imperfect national masses could be shaped into a civilized society.[33] Together, these developments built a growing consensus that

Argentina was a white, civilized, and progressing nation where indigenous peoples were an anachronistic minority destined to disappear in time due to the immutable and amoral laws of nature.

While numerous institutions and societies provided the funding, spaces, and contacts for this work, one of the most important and best studied is the Museo de La Plata. The museum came into being in 1877 when the aristocratic and self-taught collector Francisco P. Moreno organized his collections as the Anthropological and Archeological Museum of Buenos Aires. In 1884, the newly formed province of La Plata passed a law establishing the Museo de La Plata, which Moreno would build and direct for many years.[34] Building on Rebecca Earle's work, Carolyne R. Larson argues that "Through museum-based anthropological science, creole Argentines expressed a connection with indigenous cultures and bodies, simultaneously possessing and cataloguing them as artifacts belonging to the nation but not necessarily within the national community."[35] This process of *archaeologization* of the Indian or *paleontologization* of the Other allowed indigenous bodies to create cultural and monetary capital for the Argentine nation-state, even while excluding them from the benefits of citizenship.[36]

The scholars working on these topics have made clear the ways in which positivism, ethnography, anthropology, and museum culture constructed a white identity for the Argentine nation. However, although the sexual practices of "savages" fascinated nineteenth-century scholars and were believed to hold the keys to understanding human development, most historians of Latin American science and those who study science in Argentine literature have treated both the scientists and their subjects as ungendered or unproblematically masculine. In addition to almost completely ignoring women, we have only rarely considered what effect concepts such as ideal masculinity and femininity, gender roles and norms, sexual desire, and motherhood have played in Latin American anthropology.[37] Conversely, gendered histories of the nineteenth century engage with questions of race and nationalism, but usually only discuss how scientific theory impacted women, not how women impacted science.[38] The few texts that do both, such as Nancy Stepan's canonical *The Hour of Eugenics* or Julia Rodríguez's *Civilizing Argentina*, focus on the eugenics and hygiene of a later period.[39]

Just as women and sex are absent from studies of the development of racial theory in Argentina, Argentina has generally been excluded from studies of the imbrication of race, sex, and gender in colonial, imperial, and scientific contexts. Pioneering works such as those by Ann Laura Stoler, Robert Young, and Sander Gilman center on the sexual relations behind theoretical discussions of hybridity in English-language contexts.[40] Because Argentina has such a strong reputation as a white, European nation, it is often left out of these analyses. Indeed, Stoler's occasional comparisons to Latin America focus on Mexico, Cuba, and Peru. Even in *Race and Sex in Latin America*, a text geographically closer to the River Plate, Peter Wade insists that "by the late nineteenth century, Argentina already had large numbers of European immigrants, the Afro-Argentine population was small and the indigenous communities had been decimated by frontier wars. This made it easier for Argentinean elites to claim a Latin American form of whiteness." Beyond a short discussion of twentieth-century eugenics, this is Wade's only reference to the country, as if articulations of sex and race did not lead to the very conditions he claims make it less interesting to analyze.[41]

One reason for this gap in the literature is that many historians have tended to accept nineteenth-century scientific productions and first-person observations at face value, often citing the very texts they have needed to interrogate. As the only published records of creole-indigenous scientific encounters, these works become proof of "how things were" and contemporary scholars perpetuate the belief that maleness and heterosexuality were cultural norms, and thus intrinsically more valuable.[42] As a result, women are omitted from key processes of identity formation, even while their bodies were the site for the literal creation of the nation. The exclusion is even more complete for indigenous women who were marginalized for both their indigenousness and their femininity.[43] For example, Andermann's excellent text on visual and museum culture in Argentina and Brazil uses the chieftains Inacayal and Foyel as symbolic of the destinies of all indigenous peoples, but ignores the specific experiences of the half dozen women and children who accompanied them to the Museo de La Plata.[44] In this regard, he intensifies the erasures begun by Francisco Moreno and his peers who recorded the names and experiences of the chieftains, but not their wives.

This book analyzes a wide variety of materials in order to move beyond this cycle of dependence. While I engage with the work of historians, art historians, anthropologists, and others, I am, first and foremost, a literary scholar and thus read the classic anthropological texts of the nineteenth century, including Mansilla's *Una excursión a los indios ranqueles*, the complete works of Lista and Zeballos, and Moreno's *Viaje a la Patagonia austral*, from this perspective. I treat them not as objective truth but as culturally created narratives, dependent on rhetoric and literary conventions, that help us understand what Argentine elites believed or wanted to be true at a given time. By paying careful attention to the margins of these works as well as what is unsaid, we can glimpse the fractures in these ideologies. Women suddenly pop up everywhere, discussions of savagery become inseparable from anxieties about sex, and desire emerges as a motivating factor. Newspaper articles, personal letters, public records, photographs, and congressional debates enrich these emerging patterns, further permitting us to question the biases that structure the historical texts.

I also study nonfictional works in conjunction with contemporary fictional texts in order to better approximate the cultural spaces in which the Argentine anthropologists were producing knowledge. Unlike most scientists today, most of the Argentines active in the 1870s and 1880s also published popular fiction. I do not aim to prove that fictional texts were scientific, but rather to investigate why nineteenth-century anthropologists would turn to fiction, how attention to the generic conventions of fiction can illuminate the cultural bases of science, and what role these fictions played in the diffusion of scientific ideas and their related racial projects. As Sujit Sivasundaram asserts, studying science as literature and in literature permits "cross-contextualization" that allows a "more globally representative history of science to emerge."[45] Scientific texts were absolutely important for shaping concepts such as indigeneity, creoleness, and progress, but their number of readers was limited compared to those of fictional texts. In some cases, therefore, fiction played a larger role in creating racial identities than its nonfictional counterparts, even if the science itself was less precise. The inclusion of nonscientific genres also permits an exploration of how female writers who were institutionally excluded from the sciences interacted with the theories and declarations of the expanding human sciences.

Examining figures and topics that have been excluded for over a century occasionally means asking questions that may be unanswerable but highlight both indigenous peoples' intersectional relationships with creole culture and the cultural factors that dictated which questions were asked and answered in the historical record. What did indigenous peoples think about the forms of white masculinity performed for them by the explorer-scientists? Were the indigenous women that the anthropologists claimed were flirting with them doing so because they were attracted to them or as part of larger diplomatic effort? Was Ramón Lista's relationship with a Tehuelche woman voluntary or coercive? When concrete answers are not readily available, I engage in careful speculation. As Anna Brickhouse has shown in her book on indigenous mistranslation in the Americas, speculation can be a valuable methodology for diverging from a dichotomy that casts indigenous peoples as either victims or willing, romanticized participants in cross-cultural encounters.[46] In a context where we may never recuperate lost voices, it is also a mechanism for pausing to value the experiences —both positive and negative—of the oppressed.

While the topics of the texts I study vary, they were united by the authors' worldviews and supported similar ideologies. With the exception of the last chapter, the authors I study were all men born to upper-class families in the major cities of the River Plate. They identified as urban, white, educated, and well-traveled. They were further linked by a desire to study and improve their country, as well as a strong belief in progress. Even when they disagreed about details, the *porteño* (Buenos Aires) elite of the 1870s and 1880s shared a strong sense of a common goal, particularly in relation to the problems of the frontier.[47]

These authors also shared close personal connections that helped shade the similarities in their works. Lucio and Eduarda Mansilla were siblings and collaborators, as Lucio translated one of Eduarda's works from French into Spanish. Lucio Mansilla also corresponded with Francisco Moreno and served as national representative at the same time as Zeballos, who in an earlier period served with Lista's father-in-law, Olegario Andrade. Moreno was Holmberg's cousin and a good friend of Eduarda Mansilla's husband. The Argentine Scientific Society, of which Zeballos was a co-founder, funded Moreno and Lista's expeditions and provided space for them (and Holmberg) to present their research both orally and print. Juan Zorrilla de San Martín and Florence Dixie, the two non-Argentines included in this study, were on the margins of this circle but united with it by theoretical concerns. Zorrilla de San Martín was Uruguayan, Catholic, and a poet, but his epic poem *Tabaré* mused significantly on miscegenation and degeneration, inspiring Holmberg's Darwinist rebuttal, *Lin-Calél*. He also lived in Buenos Aires for several years and corresponded with many Argentine intellectuals. Florence Dixie was a Scottish woman who travelled in Patagonia, visiting many of the same places as Lista, Moreno, and Holmberg. She also hired the cook that had accompanied Lista on one of his journeys.[48] As such, the texts I analyze can be viewed as a cross section of the late nineteenth-century River Plate elite, a society where divisions between disciplines and professions were frequently mitigated by shared experience and the united opposition of the *civilizados* (civilized people) to the barbarous masses.

From Fieldwork to Fiction

The chapters in this book regender various aspects of the history of Argentine racial science, moving from observation and interactions in the field to the dissemination and appropriation of that knowledge in fiction. Throughout, I use the terms *man* and *woman* as manifestations of gender, not biological sex; that is to say, as collections of behaviors, identities, norms, and relationships that emerge from social and cultural contexts. When I refer to *sex*, I am almost always referring to the act of intercourse. It is important to note, however, that the writers I study would not have recognized such a distinction between gender and biological sex, viewing the behaviors, rights, and responsibilities they attribute to men and women as natural. Similarly, in the texts I study they conceptualize gender as a strict binary between men and women, completely eliding other forms as well as non-heterosexual sexual relations. While recognizing that other configurations surely existed in nineteenth-century Buenos Aires and on the frontier, the constraints of my textual sources lead me to work within this binary concept of gender and focus on heterosexual sex throughout this book.

Chapter One examines the classic texts of Argentine anthropology to argue that despite ostensibly focusing on describing indigenous men, the upper-class, male anthropologists centered women's bodies at the heart of their debates over racial hierarchies. In accordance with their specific beliefs about gender and sexual practices, they judged indigenous families as incompatible with the nation, in large part due to their harmful effects on women. By comparing their texts to other sources, I argue that these representations often had complicated relationships with reality. Nonetheless, the close ties between science and governing elites transformed these racial theories into gender-based disciplinary practices to hasten indigenous peoples' assimilation.

Whereas Chapter One examines white men's interpretations of indigenous sexuality, Chapter Two examines the scientists' own relationship to desire. In Mansilla's *Una excursión a los indios ranqueles,* Lista's real-life relationship with Clorinda Coile, and Lista's subsequent scientific texts, interracial sex appears as both a theoretical concern and a concrete practice. I examine how sexual discourse and action shaped Mansilla's and Lista's anthropological texts, particularly in relation to the authors' proposed relationships between the Argentine government and indigenous peoples. When read alongside my analysis in Chapter One, it becomes clear that in the last decades of the nineteenth century, interracial sex was alternatively presented as a threat to national identity and progress and a potential solution to the problems of the frontier. The fear of mixing with indigenous populations was not new, but the discourse of racial science provided a powerful new way to grapple with it.

In 1886, Francisco P. Moreno brought a group of twelve to fifteen Tehuelche Indians to live at his museum, freeing them from the military barracks where they were being held for the crime of being indigenous. While they were in the barracks and again at the museum, he arranged for their photographs to be taken. These images have come to stand as proof of Moreno's (and thus all of Argentine science's) intentions with regard to native peoples: some see Moreno as a hero and the images as a neutral historical record of different racial types; others have called Moreno a villain and the pictures mug shots of the so-called "prisoners of science."[49] In Chapter Three, I examine how aesthetic conventions, cultural codes, and the theories of racial science influenced the representation and meanings of men's and women's bodies on film. Indeed, Moreno's images illustrate a number of the relationships between women, sex, and racial science revealed in the texts studied in Chapters One and Two. This analysis also highlights further moments of potential indigenous agency that challenge the silencing of indigenous actors in many historical texts.

Chapters Four and Five deal explicitly with how literary works interact with the representations of women and sex found in more strictly scientific texts. Chapter Four looks at the ways in which fiction acted as a fertile field for debating some of the most pressing issues regarding mankind's origins, the hierarchy of races, and the advisability of mixed-race reproduction. I examine how the fictional work of Zeballos (*Callvucurá y la dinastía de los Piedra* [1884], *Painé y la dinastía de los Zorros* [1886], and *Relmú, reina de los pinares* [1888]), Zorrilla de San Martín (*Tabaré* [1888]), and Holmberg (*Lin-Calél* [1910]) used romance to popularize scientific ideas and forge the connections with a wider audience necessary to construct national identities. The sexual relationships in each text highlight the sex at the heart of debates over evolution and racial mixing, while the success or failure of differently gendered and raced pairings determined the contours of the nation, establishing who was in and who was out. These texts thus further support the argument made in Chapter One that indigenous and mestizo incorporation into the nation was not just a function of race, but also simultaneously of gender.

The anthropologists I study in this book presented natural history as a man's world, employing the tropes of the penetration of barren Argentine wilderness and the fertilizing powers of civilization. Nonetheless, many Argentine women were well aware of scientific theories and used them to make their own investigations into the place of indigenous peoples and women within the nation. In order to recuperate women's competing racial projects, I analyze the Argentine Eduarda Mansilla's 1860 short novel *Lucía Miranda* and her 1882 travelogue as well as two children's novels written in 1890 by Lady Florence Dixie, a Scotswoman who explored Patagonia, in Chapter Five. Despite the fact that they make no claims for scientific objectivity and clearly revel in sentimentalism, these texts reflect on the representations of indigenous peoples and non-indigenous Argentines produced by masculine science and policy. Through her diplomat husband, Manuel García Aguirre, Mansilla met and/or read the works of men such as Zeballos and Roca, while Dixie's two novels can be read as feminist rewritings of Musters's *At Home with the Patagonians*. Together, their works demonstrate how women carved out spaces in the debates over national identity and colonialism to examine their own subaltern position, as well how those projects affected their representations of indigenous peoples' racial fitness.

Finally, the conclusion examines the effects of the nineteenth-century nexus of scientific racism and gender on the twentieth and twenty-first centuries. I suggest that the firm break suggested by the critical literature (pre-1885 the focus is on indigenous people, post-1885 it is on eugenics and immigration) is in fact quite blurry. That is to say: scholarship has frequently treated Argentine efforts to grapple with indigenous people and immigrants as happening in two distinct historical periods with little connection between them. Despite many real differences between those two periods in the treatment of those groups and approaches to Argentine identity, in many ways the Argentine experience of assimilating natives was a proving ground for the ways of thinking that would later mediate the incorporation of immigrants. Just as many Argentine Indians continued to live, invisibly, even after the Conquest of the Desert proclaimed that they had been eliminated, the thought processes that drove Argentines to call for their exclusion continued to lurk behind policies towards immigrants and indigenous people through the twentieth and into the twenty-first century. The gendered assumptions behind nineteenth-century representations continue to take their toll: indigenous men are still represented as sexual predators in romance novels and films, and the bodies of unnamed indigenous women languish in the basement of the Museo de La Plata while their named male counterparts are slowly returned to their communities of origin. The legacy of scientific racism is powerful, and its tenets and the challenges to them will continue to shape Argentine realities for the foreseeable future.

CHAPTER 1

Inappropriate Relations

Indigenous Private Lives as a Matter of Public Concern

FROM ARGENTINE POLITICAL THEORIST Juan Bautista Alberdi's maxim that "to govern is to populate" to foundational fictions such as José Mármol's *Amalia,* the family appears repeatedly in Argentine nation-building projects of the nineteenth century.[1] Statesmen such as Alberdi and Sarmiento stressed the importance of physical reproduction, relying on heterosexual pairings to create the children that would fill the spaces of the desert. In fictional texts, the family was a synecdoche for the nation, and tales of romance did the ideological work of determining the future racial and social composition of Argentina. The family literally created future citizens and became a fertile symbolic space for negotiating belonging. In this way, the private space of the family became a matter of public concern.

Argentine anthropology shared the political and literary focus on the family. As Lucio V. Mansilla, Ramón Lista, Francisco P. Moreno, and Estanislao Zeballos traversed the frontier, they keenly observed indigenous peoples' private lives. Their upper-class creole beliefs about ideal families and sexual behavior shaped their interpretations of the cultural practices they recorded. In the anthropologists' texts, women's work, polygamy, and the kidnapping of sexual partners are offered not as curiosities, but as proof of indigenous racial inferiority.[2] Through indigenous men's action, women's bodies were thus centered at the heart of savagery as a concept.

Although these texts have generally been taken as authoritative depictions of nineteenth-century indigenous societies, by comparing their gendered representations to historical data and other works from the period, it becomes clear that they reflected racial anxieties, literary models, and the anthropologists' own performance of civilized masculinity as much as empirical observation of the places and peoples they visited.

Furthermore, through the symbolic linking of the physical family and the nation, these judgments warned of the troublesome effects indigenous family structures would have on Argentine society if they were not assimilated or corrected as the tribes came under national control. The four men's connections to governing elites, either through the sponsorship of their expeditions or through their own participation as national deputies or ministers, helped bring these private lives under public supervision. Their anthropological works laid out the justification for Argentine efforts to dismantle indigenous cultures. They also inspired several of the specific policies lawmakers implemented to do so, particularly the efforts to divide indigenous families in the 1870s and 1880s. Through these scientifically informed policy decisions, creole ideas about gender had concrete and often tragic effects on indigenous peoples' lives.

Useful Men and Republican Mothers: Gendered Expectations and the Nation

Travelers have long been fascinated by the gender roles and familial structures of the cultures they encounter. Nineteenth-century racial scientists prioritized the study of these relations, claiming, as John Lubbock did, that they provided "instructive insight into the true condition of savages" and justification for the perceived superiority of civilized societies like those in Europe.[3] The ideas the scientists brought to these encounters from their own cultures colored their observations. As theorized by present-day scholars, ethnographers in the field use abstract conceptions (experience-distant concepts) from their own cultures to understand the daily experiences (experience-near concepts) of the peoples they study, a process Clifford Geertz terms "tacking."[4] Although likely unaware of these thought processes and their implications for the accuracy of their accounts of the "true condition of savages," Mansilla, Lista, Moreno, and Zeballos applied their own upper-class creole understandings of men and women's differences, roles in society, interpersonal relations, and value to their encounters with indigenous peoples between 1870 and 1895.[5]

Like elsewhere, Argentine understandings of gender roles shifted over time and in response to pressures including religion, class, politics, and demographic transitions.[6] The writers I study, like other well-to-do, urban Creoles of the time, imagined the nuclear Christian family as the basis of the Argentine nation. Their understandings of the family and men and women's roles within it shared many similarities with Victorian culture. Families were composed of a husband and his wife, and responsibilities were clearly defined by gender. The ideal woman was the Angel in the House, responsible for protecting the home's spiritual center, a connection strengthened by Hispanic Catholicism's veneration of the Virgin Mary.[7] She was also responsible for domestic tasks and birthing and raising children. Of course, many women could not or did not want to meet this ideal.[8] Nonetheless, it remained in circulation among Argentine elites throughout the late nineteenth century, profoundly influencing large portions of the nation through upper-class Buenos Aires's influence on politics, law, literature, science, and related fields.

Men, in contrast, moved freely through public and private spaces.[9] This division of both space and power was strengthened by the 1869 Argentine Civil Code (effective 1871) which clarified the rights of the masculine head of household (*patria potestad*). It reiterated a husband's authority over his wife and his responsibility to provide for her by classifying both women and children as minors.[10] It also required a woman's husband's permission in order for her to enter legal action; buy, sell, or mortgage property; become employed; or administer her own wages.[11] It further reduced her legal rights over her children and required her to obey her husband's wishes, including with regard to her place of residence.[12] And although not defined by the Civil Code, a series of Supreme Court cases in the same period ruled that a woman's citizenship followed her husband's, thus effectively expatriating any woman who married a foreigner.[13]

Both men's and women's responsibilities were conceptualized as exceeding intimate spaces in order to provide essential services to the nation. Good republican mothers educated their offspring to be productive future citizens.[14] Juan Bautista Alberdi made this connection clear in his influential treatise, *Bases y puntos de partida para la organización política de la República Argentina* (1852), when he argued that a woman "organizes the family, prepares the citizen, lays the foundations of the State."[15] Masculine power and productivity were similarly tied to the success of the nation. Just as men were expected to financially provide for their wives and children, their public activities were also supposed to contribute to the common good. This expectation was often expressed through the concept of the *hombre útil*, or useful man.[16] Both men and women's activities thus had consequences for the individual home as well as the nation and the *raza* (race) that structured it.[17]

Argentine scientists drew upon these beliefs when studying indigenous peoples in the Pampas and Patagonia. Confident in their own representative status with regard to their race, they largely focused on observing and judging indigenous men. In the antebellum period in the United States, women were nearly absent as subjects of ethnographic inquiry and "race in this literature was largely a matter of male descent."[18] Similar patterns are seen in the Argentine literature: when it came to describing the characteristics of different indigenous societies and evaluating their potential to be civilized, women and children mattered primarily in relation to men, particularly with regard to how the men treated them.

Although the explorers studied different tribes and had varied attitudes towards the people they encountered, their shared cultural background and readings led to two dominant images of indigenous manhood. The first was characterized by a laziness and dependence on women that implied an abdication of the role of family patriarch. The second trope was of the hyper-masculine Indian, an irrational, violent, and excessively sexual being. Both characterized indigenous masculinity as threatening to the formation of the ideal creole family and thus excluded indigenous peoples in their present state from the nation. In both cases, masculinity and its relationship to savagery cannot be conceived of separately from women and their bodies.

The Threat of Inadequate Indigenous Masculinity

Nineteenth-century and more recent sources agree that among most of the central and southern Argentine indigenous communities, men's primary occupation was hunting. This hunting often happened cyclically, meaning that men's activity level shifted according to seasons, the availability of game, and the needs of the community. Furthermore, by the mid-nineteenth century the spread of cattle meant that men were no longer obligated to continually chase guanacos and ostriches, freeing up their time.[19] In contrast, women were largely responsible for all other tasks, including child-rearing, weaving, cooking, cleaning, organizing moves from one place to another, and setting up camp. Many of these tasks were difficult and daily.

To the creole scientists who believed in and personally embodied the ideal of productive masculinity, these social structures were both foreign and improper. Their descriptions of them are heavily laden with value judgments. From the Guaraníes of the north to the Onas of Tierra del Fuego, there is surprising unity in the Argentine anthropologists' descriptions of indigenous men as indolent to the point of absurdity. *Haragán*, *perezoso*, and *apático* (idle, lazy, and apathetic) are among the most common words used to describe the various tribes, found in every text I study. In *Viaje a la Patagonia austral*, which recounts his 1876–77 excursion, Moreno described the Manzanero male as "an idler like all savages," spending his days lying face down or relaxing on a *quillango* (fur blanket).[20] Indeed, Lista recounted, the Tehuelche men were so lazy they would go days without eating just to avoid the exertion of searching for food.[21] Both comments clearly take a tone of reproach regarding indigenous men's action, or lack thereof.

That men's limited responsibility to the family should be interpreted as a negative characteristic is highlighted by the implicit contrast with the anthropologists' own performance of civilized masculinity. As Claudia Torre notes in her study of texts from the Conquest of the Desert, the Argentine explorers narrated their activities within the framework of a "vehemencia del hacer" ("passion for doing") that exemplified the creole association of masculinity and public productivity.[22] Mansilla wrote of the great pleasure he found in the "manly exercise" of crisscrossing the pampas.[23] All self consciously depicted themselves continuously traveling, thinking, and producing knowledge and artefacts for the consumption of the Argentine elite. Even the structure of their texts supported this performance: most of their ethnographies take the form of travelogues that privilege the scientists' purposeful movement across time and space. And although the men do not reference their own families, their paternalistic attitudes towards indigenous peoples as a whole and the godfatherships they establish with particular individuals allow them to meet those expectations by subbing homeland for home.

The laziness of indigenous men threw in doubt their ability to be active members of the emerging nation. Indigenous men's failure to be productive and provide for the family meant that they were not real men, but rather "true children," to use Lista's words.[24] This characterization extended beyond the nuclear family, for as children, they would also fail to provide for the national family, condemning them to a perpetual state of inferiority and dependence on the Argentine state. This relationship is foreshadowed as the explorers—civilized patriarchs and official representatives of Argentina—found themselves repeatedly barraged with requests for alcohol, sugar, and meat. Rather than work or farm as nineteenth-century creole norms dictated men should, the indigenous men depended on national authorities for survival.[25] In this trope, so present in the Argentine anthropological literature, the *hijos del desierto* (children of the desert, a common Argentine term for indigenous peoples) were truly childlike, begging for Argentina to take control and save them from themselves. As Moreno wrote, "These poor indigenes, in their relationships with white men, have truly infantile appearances."[26] These judgements are also visible in public discourse from the period: an article in *La Prensa* in December 1878 recommended that subdued indigenous adults be legally classified as minors and be given creole guardians to help them adapt to civilized life.[27]

For the anthropologists, indigenous men's laziness was also problematic because of its effects on women. In fact, one of the primary techniques Moreno, Lista, and Zeballos used to characterize the Indian male as excessively lazy was explicitly contrasting his lack of activity with the industriousness of the indigenous women they encountered. Each author described women taking care of children, preparing food, making clothing, planting, packing up camp, and otherwise ensuring the survival of the family.[28] Whereas the anthropologists most frequently described men as *apáticos* (apathetic) or *indolentes* (indolent), they used the adjective *hacendosas* (industrious) to characterize indigenous women.[29] In one of the most powerful descriptions of this gendered juxtaposition, Lista reproachfully reported that Ona men devoted themselves to lolling about in their huts while the women stripped down to gather shellfish in freezing austral waters.[30] There is no reason to believe these descriptions of men and women's behaviors were factually incorrect, but it is clear that Lista, Moreno, and Zeballos's cultural paradigms shaded their interpretations of those behaviors and thus the tone and purpose of their narrations.

Ironically, like indigenous men's unproductivity, women's productivity was evidence of indigenous families' inability to contribute positively to the nation. Zeballos, Lista, Moreno, and Mansilla depict indigenous women's participation in what would be considered masculine activities in the creole world as always involuntary, and thus a sign of savagery. For example, Lista claimed that the Indian woman was the victim of a lazy master, "sentenced to perpetually drag the heavy chain of slavery."[31] She was undervalued, treated as property, and subject to be killed when times were lean and food scarce.[32] As Mansilla summed up in his 1870 ethnography of the Ranquel Indians, "In those lands, women have only two purposes in life: to work and to bear children."[33] From their perspective, women did not chose to be productive; they were productive because their savage cultures had not yet developed the respect for women that characterized civilization.

That achievement of creole masculine ideals, particularly providing for women, was an important condition of civilization and thus the ability to be incorporated into the Argentine nation is proven by the few exceptional tribes who more closely approximated creole norms. In *La conquista de quince mil leguas*, which directly influenced military actions in the Pampas, Zeballos claims that the Tehuelche Indians are "apt for civilization and to serve as assistants in the colonization of those territories," for their society is patriarchal and they respect "the family and their social order."[34] In his novel *Relmú*, he repeats these connections to describe the Picunche as an ideal indigenous society. Instead of the "traditional laziness of the Pampa horsemen," they love work, value a home life, and "almost all dedicated themselves to agriculture."[35] In these passages, masculine productivity and respect for the patriarchal, nuclear family are signs that certain tribes will be better equipped for incorporation to the nation. In the case of Zeballos's texts and their appropriation by military authorities, these characteristics may very well have saved certain tribes from being physically eliminated by government forces.

Were indigenous women in fact powerless slaves, victimized by barbaric, lazy men? On the one hand, it is impossible to deny that life for many indigenous women was difficult, and that their responsibilities were constant. Nonetheless, as Rebecca Jager has shown, across the Americas white ethnographers' narrow definitions of femininity and female roles meant that they often ignored or misread indigenous women's important contributions to "subsistence, commerce, international relations, and the spiritual and social organization of their communities."[36] Work that granted women cultural, political, or economic power was either overlooked because of white ethnographers' gendered expectations or classified as drudgery and thus proof of the lowness of the race.

Various contemporary and present-day accounts suggest that Jager's hypothesis holds true for Argentina and that women did hold some power within Argentine indigenous societies. The female chieftain María dominated Patagonia for much of the early nineteenth century, negotiating with other tribes and outsiders from Argentina and England.[37] Other women participated in war as soldiers and scouts.[38] Among the Tehuelches, first-born daughters, the wives of chieftains, and other female elites enjoyed certain privileges denied to men of lower social classes.[39] Furthermore, Miguel Angel Palermo has demonstrated that women in northern Patagonia (explored by Lista, Moreno, and Zeballos) exerted near-total control over the textile industry. This work meant that they were deeply involved in commerce, ceremonial rituals related to diplomacy, and interpretation/translation, among other activities.[40]

Moreno himself describes negotiating with a Gennacken woman named María, the wife of a chieftain. She had come to him to trade and became his entrance into the indigenous world, obtaining permission for him to travel, supplying his team with food, and negotiating with other Indians on his behalf.[41] Several days later, María is the only person in the camp with a horse available to rent to Moreno. Desperate for a mount, Moreno must negotiate with her, supplying her with gifts of sugar and other products as well as giving in to her demand that he ask her favorite dog for permission to borrow "his" horse with a completely straight face.[42] This passage demonstrates not only María's wealth and power within her tribe, but also creole explorers' dependency on the very women they dismiss as powerless slaves.

Women's work may have been tiresome, but as the above examples show, it also provided them with some degree of economic standing and power. Like women's activities, this potential power collided with the expectations of elite Argentines, particularly the concept of *patria potestad* elaborated in the Civil Code. Consequently, Lista, Zeballos, Mansilla, and Moreno recast indigenous women's activity as evidence of racial inferiority defined by masculine failure, a failure that paved the way for forced "correction" of these behaviors in order to bring indigenous families in line with creole ideals.

The Threat of Excessive Indigenous Masculinity

In stark contrast to the image of indigenous men as children who fail to protect women, a second representation emerged of indigenous men as excessively masculine. Later in *Viaje a la Patagonia austral*, Moreno hints at this alternative masculinity when he repeats the image of the childlike Indian with a caveat: "they are morally children (except for their vices, which are those of men)."[43] Moreno does not clarify what these vices are, but his texts as well as those of his peers made clear that indigenous men could also be too masculine because of a sexuality that was excessive, irrational, bestial, and especially threatening to white women. In this case, the harm to women caused by indigenous men's failure to act was replaced with harm caused by direct sexual action.

As in the case of the gendered division of labor, indigenous sexual practices differed from creole expectations. First, as a number of the anthropologists noted, polygamy was a common practice, although it was almost always restricted to chieftains and other elites.[44] The scientists also called attention to indigenous marriage practices and women's sexual freedom, which went against the recommendations of the Catholic church and the state. Finally, several of the anthropologists recorded that many indigenous groups relied on the kidnapping of women as both a wartime strategy and a way to announce one's intentions to marry.[45]

These practices were similarly subject to value judgements deriving from the gaps between creole ideals and indigenous customs. In *Descripción amena*, Zeballos notes the "astonishing virility" of the Araucanos.[46] This virility is not associated with the positive rationality of civilized masculinity, however, but rather a negative excess of destructive sexual energy that served to break things down instead of building them up. In *Viaje a la Patagonia austral*, Moreno writes of observing "the most atrocious orgy" motivated by "horrific drunkenness, the most disgusting debauchery." The scene is hellish, marked by nakedness, blood, and violence. He notes that the Indians become more lecherous when given alcohol and compares them to monsters.[47] As in the descriptions of women's work, the sexual excess and violence of indigenous men are described in moralizing terms, establishing "links between deviant sexuality and savagery."[48]

The scientists depicted this sexual excess as driven by inherent racial characteristics, augmented by alcohol, and oppositional to rational masculinity and thus more closely related to animal habits. Mansilla claimed that indigenous men "stalked" their victims while under the spell of "erotic furies," highlighting their irrationality and bestiality.[49] Zeballos called the Indian male a "frenzied barbarian," again characterized by an excess of masculine energy that is "inflamed at times by the impulse of uncontrollable passions, as is the desert hurricane itself."[50] The metaphor of the hurricane, like Mansilla's "stalking," emphasizes the inhuman nature of indigenous sexuality, as does Zeballos's assertion that indigenous men were like starving wolves.[51] The dehumanizing of the Indian male continued with Zeballos's chosen epithet, "the American centaur," a nickname that references both the American Indian's ever-present horse and the lasciviousness Greek mythology associated with the beast.

In his fictional trilogy, closely based on nonfictional texts, Zeballos developed the trope of the hypersexual Indian to maximum effect. The characterization of indigenous men as lustful is established clearly in the second book, as the eponymous chieftain, Painé, is repeatedly described as "lecherous."[52] The narrator, Liberato, asserts that the chieftain's lust "did not recognize boundaries," and as a result he had several wives as well as multiple captive lovers.[53] This carnality is not unique to Painé. Liberato attributes a war between two Ranquel chieftains to a romantic entanglement caused by the "insatiable boil of lust" of an old cacique for a young Indian girl, only thirteen years old.[54] Again, excess and irrationality define indigenous sexuality, and native diplomacy and politics are reduced to the unreasonable desire of the chieftain, impotent yet lusting after a newly pubescent girl. In each of these examples, indigenous women are cast as innocent victims engaged in a sort of sexual slavery that matched the drudgery of their daily work.

This masculinity was largely characterized by the threat it presented to white, creole women. During his eighteen-day stay among the Ranquel Indians, Mansilla encountered numerous captive white women, as well as several mestizo caciques whose very existence pointed to prior relations between white captives and their Indian captors. Faced with the reality of mixed-race relations on the frontier, Mansilla attributes this phenomenon to indigenous men's violent lust for white women. In *Una excursión*, he pities the women who must suffer the "erotic furies of their master" and the jealousy of the Indian women, painting captivity as a hellish period of physical and sexual abuse at the hands of predatory animals.[55] Zeballos also repeatedly wrote of how the Indian male trampled across civilized society, leaving "tracks stained with blood" and a path of tortured captives in his wake.[56] In the chronicle *Callvucurá*, Zeballos establishes the Indian's desire for white women when the eponymous chieftain promises his warriors lands "rich in cattle and Christian women."[57] Creole women are equated with livestock, booty to be stolen, and indigenous men display none of the respect that upper-class Argentine society purported to have for good Christian women.

As Gabriella Nouzeilles has shown, this image was a wild affront to the reasoned white masculinity of the scientist-explorers and their national projects.[58] For these men, conditioned by their own upper-class gender norms, faith in order, and anthropological understandings of human progress, unbridled masculinity was an important marker of the indigenous communities' inferior status. In addition to threatening the integrity of Argentine women and, through them, the honor of men, the hypersexualized Indian male also represented a threat to the nation-family by taking away the civic mother. Given the real and symbolic power associated with motherhood, if she were taken captive, her children would be left without a mother, and her home lose its spiritual and moral center. The good Christian family was thus broken, threatening to reduce the foundation of the nation to a pile of rubble. Who would educate the children? Who would protect the nation's morality and tradition? Given that Argentine intellectuals repeatedly posited the family as the building block of the nation, the interruption of Christian, nuclear, civilized pairings and their replacement with socially and racially unbalanced relations meant nothing less than the subversion of the Argentine Republic. Through this dynamic, the hypermasculine trope, like its negative, failed masculinity, postulated that indigenous men were a serious problem that needed to be resolved.

The hypersexual trope also resolved another issue for writers of the Argentine frontier: how to deal with very real mixed-race couples while simultaneously asserting the fundamentally white identity of the nation. Mixed-race children born on the frontier were useful fodder for European meditations on hybridity, but presented a serious problem for "a society [Argentina] whose national project has been to achieve whiteness."[59] The captive woman thus became a site where all the nation's anxieties over racial mixture could concentrate.[60] Zeballos and Mansilla insisted on the righteousness of the white female captive, who they nearly always represent as a virgin or a mother sobbing for her murdered children.[61] Both further emphasized that the captives were not attracted to their captors. Mansilla expressed great admiration for the handful of "heroic women" he met who had chosen to be beaten and abused instead of giving in to the sexual demands of their captors.[62] By depicting captive/Indian pairings as the result of the unlimited violence of indigenous men, any possibility of women choosing to take indigenous husbands or lovers is elided. White women could thus not be blamed for the intentional subversion of Argentine whiteness, nor did the Argentine men have to admit that a woman might prefer a so-called savage over civilized life.

There are good reasons to believe that this hypermasculine trope responded to upper-class creole anxieties about sex and race as well as to real-life sexual practices. Whereas the image of the lazy Indian can be found in works from around the world, the hypermasculine trope is nearly completely absent from corresponding European anthropological texts. Rather, the American Indian had a long history of being represented as effeminate. Eighteenth-century scholars such as William Robertson and the Comte de Buffon believed one of the first signs of the Indian male's undersexed nature was his beardlessness.[63] In the one of the most extreme versions of this trope, Cornelius de Pauw argued that American Indian men were even capable of producing breast milk.[64] This vision lasted into the nineteenth century. Although Europeans from that period complained of the American Indian's lack of true love, disregard for chastity, short marriages, and wife exchange, this sexual deviance was circumscribed within the tribe and not presented as threatening to the nonindigenous population. This representation contrasts sharply with that found in many of the Argentine texts.

Additionally, while it is true that the captivity of white women and children was unfortunately common along the contested frontiers of the nation, historical accounts suggest that, overall, the rape of female captives in Argentina was rarer than urban elites suggested. Data from the 1830s show that the fertility rate of captive women was low: a captive woman had a 7 percent chance of having a living child in any particular year. Socolow suggests that one reason for this reduced rate was that although captive women married indigenous men, they "were not particularly attractive as sexual partners."[65] Similarly, Rotker argues that, while in later decades there were often thirty to fifty captives per tribe and captive women were forced to endure hard labor, the majority of accounts agree that "nothing indicates that the captives were abused sexually. It may be that many were, but their consent seems to have been necessary on the whole."[66] Finally, analysis of legal and baptismal records from the nineteenth century demonstrates that contrary to what Mansilla and Zeballos claimed, there were numerous voluntary mixed-race couplings on the frontier.[67] According to Argeri, interracial couples in contact areas even relied on kidnapping-like behaviors in order to overcome the opposition of relatives.[68] Although these practices looked similar to involuntary captivity, both the man and woman would be acting of their own free will. In this regard, the anthropologists' depictions of captivity make visible the gap between urban upper-class racial aspirations and the practices of lower-class residents in rural spaces.

Comparing Zeballos's texts with their sources further highlights the discrepancy between reality and his representation. Often referred to as an armchair ethnographer, his first-hand experience on the frontier was both shorter and later than that of most of his peers. Although chosen as an expert on indigenous history and anthropology prior to the Conquest of the Desert, Zeballos did not actually visit Patagonia himself until late 1879. At this point he had already published his influential *La conquista de quince mil leguas* and the Roca campaign had begun to alter the human and ecological landscape of the region. As such, Zeballos relied heavily on written texts, including a set of manuscripts he claimed to have found in the desert in 1879.[69] Those manuscripts, written around 1854 and later published as *Memorias del ex-cautivo Santiago Avedaño* and *Usos y costumbres de los indios de las pampas*, are a first-person narration of the Argentine Santiago Avedaño's eight-year captivity among the Ranquel Indians.[70]

Zeballos's texts' accentuation of the dangerous, animal sexuality of indigenous men is dramatically different from the representation found in Avedaño's manuscripts, Zeballos's primary source. The rape of captive white women does not appear at all in Avedaño's recollections. He does mention female captivity in a few passages, but like the European scientists, he inscribes it within the indigenous world. In the vast majority of captivity episodes Avedaño mentions, the captives are from other indigenous tribes. They are not white women. His tales also do not suggest frenzied sexual assault.[71] Given Avedaño's eyewitness status, it stands to reason that if women were in fact suffering the advances of barbarous predators, it would appear in his account.

In fact, Avedaño directly refutes the charges of sexual cruelty to white women, exploiting the authority of personal experience to make his point. He addresses the horrifying claims of the abuse of women by indigenous men in the following fashion: "It is impossible for me to believe these things that they say, or that these atrocious incidents have happened, without me having found out, not even through news, given that I stayed eight years and nearly eight months with the Ranquel Indians, from 1842 to 1849."[72] Avedaño also makes clear that many women on the frontier came to enjoy their indigenous husbands and refused to return to white society so as not to leave behind their mixed-race children.[73] In these descriptions, Avedaño simultaneously undermines the Indian-as-rapist trope and allows white women some degree of sexual agency. Zeballos's depiction of the hypersexualized male is thus entirely his own addition to Avedaño's narrative, part of his general tendency to elide passages in Avedaño that depicted indigenous people in a positive light and overemphasize those that depicted them as cruel, savage, or inferior.[74]

Indeed, Zeballos's dramatic depictions appear to have much more in common with classic fictional Argentine literature—including Esteban Echeverría's *La cautiva* and multiple versions of the Lucía Miranda myth (including Eduarda Mansilla's 1860 novel, studied here in Chapter Five)—than with any sort of reality.[75] While some captive creole women were most certainly sexually assaulted, elite male anthropologists like Zeballos appear to have amplified this experience in the literature in order to create a monolithic enemy against which to construct the nation and to justify military actions against the diverse indigenous populations who occupied economically valuable lands.[76] These representations often put them at odds with the practices of the lower-class residents of the frontier, driving home the point that scientific concepts were often deployed by elites to serve urban, upper-class projects.

The stereotype of indigenous laziness and that of the indigenous sexual predator emerged from assumptions and anxieties about gender roles, reflecting what Argentine anthropologists believed or needed to be true. Whether too masculine or not masculine enough, indigenous men posed a threat to the imagined ideal creole family and thus were incompatible with the rise of the Argentine Republic. Although only partially rooted in reality, these tropes have persisted for over a hundred years. The anthropologists' texts continue to be the principal sources for histories of the

Argentine nineteenth century. Some, including Mansilla's *Una excursión a los indios ranqueles* and Zeballos's *La conquista de quince mil leguas*, are part of grade school curriculums in Argentina and graduate curriculums both in the country and abroad. Additionally, predatory and/or lazy indigenous men have appeared in a wide variety of popular cultural productions. *La Vuelta del Malón* (1892), Angel Della Valle's painting of a naked white woman carried on horseback by a returning indigenous raider, is likely the most well-known Argentine artwork of all time. In 2010, it even appeared on the cover of a bestselling Argentine romance novel, Florencia Bonelli's *Indias blancas: La vuelta del ranquel.*[77] Most tragically, the images of inappropriate masculinity developed in the anthropologists' texts lay the bases for the development of forced assimilation policies as well as the imprisonment and murder of indigenous peoples. As the next section will show, just as gendered failures were an important indicator of indigenous inferiority, policies meant to "improve" the tribes took aim at indigenous families, creating different destinies for men, women, and children.

Anthropology Made Policy: Disciplining the Indigenous Family

The Argentine science of indigenous gender roles and familial structures reached intellectual and political elites through newspapers, academic journals, scientifically inspired fictional texts, seminars, and lectures promoted by organizations like the Argentine Scientific Society. The gendered representation of indigenous men and women and their inadequate familial structures also landed directly in the hands of policy makers through government sponsorship of scientific activities. Moreno's *Viaje a la Patagonia austral* and most of Zeballos's and Lista's texts were paid for by the Argentine government.[78] The ethnography in Mansilla's *Una excursión* was the result of a journey undertaken in order to ratify a peace treaty between the national government and the Ranquel Indians, and General Julio Argentino Roca commissioned Zeballos's *La conquista de quince mil leguas* as a blueprint for the 1879 Expedition to Río Negro.[79]

The four anthropologists studied here also played a direct role in Argentine government and policymaking. In addition to his military positions, Mansilla was governor of the Chaco Territory (1878–1880) and a representative in the National Congress (1876, 1885–1891). Lista was governor of Santa Cruz province from 1887 to 1892. Zeballos was elected to the National Congress in 1880, 1884, and 1912, and served as Minister of Foreign Affairs on three separate occasions (1889–1890, 1891–1892, 1906–1908). The only one of the four not to serve in a legislative role during this time period, Moreno was the director of the provincially funded Museo de La Plata and an official *perito* (expert) for the Argentine national government. In the case of nineteenth-century Argentina, science truly did not exist separately from politics, and it is thus necessary to follow scientific thinking through observation and theory into its real influence on governmental policy.

Just as the anthropologists I study found men's treatment of women to be key for assessing different tribes' state of civilization, the family was quickly targeted as a productive arena for eliminating signs of savagery from those culturally othered citizens. Even as they invalidated existing indigenous family structures, Argentine elites recognized the family's importance in the preservation of language and culture and thus the continued threat it posed to creole hegemony over the national territory and racial identity. Disrupting indigenous families or changing gender roles within them could encourage Argentine Indians to assimilate to hegemonic ways of life.

Polygamy quickly became a target of Argentine efforts to integrate indigenous peoples. In an order to his troops on April 30, 1879, General Julio A. Roca encouraged military leaders who associated with Indians "to exert utmost caution to assure that they conform to the good manners of civilization, absolutely prohibiting marriage with two or more women and other tribal ceremonies that in any way offend morals and decency." He encouraged his men to use "repressive measures" if they deemed them necessary for achieving that goal.[80] As this episode makes clear, it was not enough for indigenous peoples to be under the physical power of the Argentine government. The particular groups he is referring to were Indians already under control as soldiers, allies, or prisoners, and who thus did not pose a physical threat to residents of the territory. True domination would only come by also regulating their private lives, particularly their sexual practices and family structures.

Later that year, Friar Pio Bentivoglio wrote to Friar Moysés Alvarez in order to update him regarding his activities on the frontier at Fort Sarmiento. His letter shows that Roca's order had spread, for he explains that Coronel Racedo had charged him with dealing with petitions from the "friendly Indians" asking for women to help with cooking and cleaning. He was to make sure that Roca's order was enforced and that "moral principles prevailed." In order to do so without "clashing with the animal demands [of the Indians]," Bentivoglio claims he always sent them the ugliest and oldest women, perfectly able to cook and clean but undesirable as sexual partners.[81] In this way, Argentina's ruling class continued to push indigenous peoples toward assimilation by "correcting" their sexual practices and restricting possibilities for reproduction.

As he made clear in an 1878 editorial in *La Prensa*, Roca also was a major proponent of dividing indigenous families as a complement to military actions in the Conquest of the Desert.[82] The following year, he sent an Argentine military official to the United States to observe US indigenous policy. That envoy, Miguel Malarín, focused on gendered activities as a way to put an end to the "eternal national nightmare" that was the *cuestión de indios*.[83] He approvingly noted that in the United States, tribal dispersal had resulted in a lower birth rate and that the Trail of Tears had caused women to miscarry and children to die, outcomes he viewed as sad but necessary. His analysis of US history suggested that another option for limiting reproduction was alcohol, which "shortens the lifespan and sterilizes women." Although he judged them effective, in the end Malarín dismissed these options as too slow to cause a real transition on their own. Instead, he recommended they be combined with forced work and the development of indigenous colonies.[84]

The opportunity to put these theoretical concerns into practice came in the late 1870s and early 1880s as military columns rolled across the pampas and large numbers of indigenous people came under the control of the Argentine government. While popular imaginings of the Conquest of the Desert focus on physical extermination, far more indigenous people were imprisoned than killed. The arrival of thousands of captive indigenous men, women, and children to Argentine cities fundamentally changed the tone of the debates over the *cuestión de indios.* The antagonistic binary "civilization or barbarism" was cast aside as intellectual and political elites began to conceptualize "civilization and barbarism," or how to best ensure the assimilation of the defeated tribes to creole culture.[85]

Several historians have investigated what happened to subdued Argentine Indians during and immediately after the Conquest of the Desert. After encounters with the military, indigenous peoples who had not been killed were taken prisoner and shipped to Buenos Aires or other port cities. In Buenos Aires, many were housed on Martín García Island in a setup some scholars have described as a concentration camp.[86] Living conditions were poor, resources were lacking, and diseases such as smallpox ran rampant.[87] From there, indigenous men were conscripted into the military or sent to work on plantations in the northern provinces of Tucumán or Entre Ríos, far from their homes in the south. Women and children were distributed as domestic help, either to elite families who requested them or through a sort of public auction organized by the Sociedad de Beneficencia.

In many ways, the distribution system responded to the practical needs of the growing nation. Subdued men were a cheap labor source and filled holes in the front lines of the Argentine army and navy. Women were useful as household help, and children could be trained at a young age to contribute in gender-appropriate ways to the economy. Most importantly, distribution was a way to transfer the financial and logistical responsibility of maintaining indigenous peoples from the central government to other entities.

Beyond these pragmatic justifications, the distribution system also disciplined gendered indigenous bodies and dismantled the indigenous family's reproductive capability in order to hasten the cultural destruction of the Argentine Indian. In anthropological writing, indigenous men were maligned as useless and overly sexual; the distribution system worked to correct both perceived faults. Throughout the Americas, there existed a belief that work defined masculinity; to be male was to be productive.[88] Contracting indigenous men out to plantations would teach them to work the very ground they were accused of leaving fallow, fulfilling Zeballos's recommendation that the only way to resolve the problems of the interior was to "Take away the Pampa Indians' horse and spear and obligate them to work the land, with a Remington to the chest daily."[89]

Indigenous men were also incorporated into the army and navy, continuing a long Argentine history of using the military to discipline unruly men. Juan Manuel de Rosas's regime (1829–1832 and 1835–1852) used forced conscription as punishment for crimes, real and invented, especially among the rural lower classes.[90] Redistribution of pacified Indians followed this example, and by 1881 indigenous men composed approximately one third of military units.[91] Within the armed forces, recruits were taught to work and socialized in proper masculinity as laziness and excess ceded to measured obedience. Contemporary observers were shocked at how quickly the Indians became "disciplined and efficient recruits," working for the nation instead of against it.[92] In sharp contrast to the indolent *hijos del desierto*, on Argentine ships the "regenerated Indians," boasting new names like Bismark and Garibaldi, "r[a]n happily to carry out with great precision the maneuver with which they had been tasked."[93]

The distribution of Indian males to almost exclusively masculine institutions also curbed the productive and destructive aspects of indigenous sexuality by imposing celibacy. Segregated in barracks or on naval vessels, they no longer presented a (real or imagined) threat to white women. Access was physically cut off in a way that the frontier had been unable to ensure. Gendered isolation also separated indigenous couples, reducing heterosexual sexual contact and limiting the creation of future Indians.[94] This policy thus worked to both culturally and physically diminish indigenous presence in the nation.

Women were distributed to wealthy Argentine families and trained by other women as domestic servants, similarly undergoing a process of reeducation in creole gender norms. While this work was important, elites were most concerned with indigenous women in their capacity as mothers. Since many believed that mothers shaped future generations and ensured the propagation of culture, indigenous mothers threatened to raise a new cohort of barbarians who would extend Argentina's struggle for control over its territories and civilians. Consequently, Roca's envoy Malarín recommended that "The little Indians should be divided up among the families of the Republic, with certain obligations to them. It is not the old encomienda system, rather tutelage until they come of age in order to civilize the savage."[95] Malarín's recommendation was likely inspired by his experience in the United States, as the late nineteenth century saw a surge in programs designed to improve indigenous children by forcibly removing them from their families.[96]

Military leaders were not the only ones recommending this separation. In 1877, the Superior de las Misiones a los Indios, Father Pablo Emilio Savino, asked for a monthly sum of five thousand pesos in order to found two boarding schools for indigenous children, one for each gender, as a civilizing counterpart to military actions. These schools would train both sexes in hard work, teach them "skills and trades," and train the girls to be nuns who could teach other indigenous populations.[97] Savino insisted that although slower, starting with children would pay off in the long run. A few decades later, both the Salesian missionaries in Patagonia and the Franciscans in Northern Chaco also argued for boarding schools by explaining that separating children from their parents would yield excellent results in just a few years.[98] The replacement of bad indigenous mothers with the motherly teachings of the Catholic Church would take advantage of the power of the family to bring about total societal change. Like Savino, both the Salesians and Franciscans proposed distinct education plans for boys and girls, demonstrating once again that the transformation towards civilization was a gendered project.

While few boarding school projects reached completion, the distribution system served as an effective knife for separating indigenous children from their mothers. Several requests for Indians ask for specific ages and make no reference to the children's parents, such as Nicasio Oroño's 1879 request for "a pair of *chinitas* [little Indian girls] about six or eight years old."[99] Other indigenous people were scattered via public *reparto* (distribution) organized by the upper-class ladies of the Sociedad de Beneficencia. According to Emilie Daireaux, a contemporary French observer and good friend of Moreno, the reparto frequently resulted in the disruption of indigenous families:

> At the designated hour, the herd was brought there and settled without brutality, but also without compassion. One could see poor old women, whom no one would want, with their gray and lank hair; young women who breast-

> fed, or gathered their numerous children around them; and wayward young boys and girls already separated from their mothers, whom they had lost in the revolts and disruptions of the desert and in the disorder of the embarkations. . . . The citizens would come up and carry off, out of charity or self-interest, a few of those beings separated from humanity. One could observe heart-wrenching scenes that no one paid attention to; some mothers gripped their daughters, but they were rejected.[100]

As Daireaux's account reveals, some separations were intentional, giving rise to horrific images of women desperately clinging to their children to no avail. Other separations were the result of negligence on the part of the authorities. We can imagine that many of the cases of "wayward young boys and girls" could have been avoided. Once the *indiecitos* (little Indians) were doled out, they were often baptized by their adoptive families and given Spanish names, symbolizing their new life as Christians. Because the church was in charge of all registries until 1884, baptismal certificates often functioned as identity documents. When the Indians were baptized, the Creole in charge of them would receive the *fe de bautismo* (baptismal certificate) as well as *derechos de potestad* (parental authority), stripping the child's biological indigenous parents of that legal right.[101] These adoptions thus involved a symbolic, practical, and legal destruction of the ties between parents and children in indigenous families. The important role of the entirely female Sociedad de Beneficencia in this process underscores that women were active participants in constructing national identity, including in the physical reshaping usually associated with masculine military efforts.

Although the exact number of indigenous children affected by these policies is unknown, Bustos and Dam's study of the town of Carmen de Patagones provides a glimpse into the extent of these separations. They analyzed the *Registro de Vecindad* (residence registry) and found that of 2,733 residents in 1887, 130 were labeled *indio*, *india*, or *china*, all terms used to describe native populations. Of those 130, 104 were children, and 97 of them lived with creole or immigrant families.[102] Thus, an astonishing 93 percent of indigenous children in Carmen de Patagones lived in nonindigenous households. Many of those children were listed in parochial registers as "hijos de padres indios desconocidos" (children of unknown Indian parents), providing support for the supposition that they lived apart from their biological families.[103] Similarly, in *Estado y cuestión indígena*, Hugo Mases identifies 475 Indians, mainly women and children, who were distributed in Buenos Aires. The vast majority were under eighteen years of age. Even some of the youngest on Mases's list, including two-month-old Lorenza

Elisa Pizarro, are listed as living with creole parents and no indigenous adults.[104] In both cases, the specific circumstances of the children's separation from their parents are unknown, but it is probable that many, if not most, were involuntary.

These policies were originally justified for the public in newspapers such as *El Nacional*, headed by former president Domingo F. Sarmiento. On November 30, 1878, an editorial in that paper argued that the separation of families was for the benefit of indigenous children. Keeping children older than ten with their families, the author claimed, "is to perpetuate the child's barbarism, ignorance, and ineptitude, condemning him to receive moral and religious lessons from the savage woman. There is charity in separating them as soon as possible from this ruin."[105] From this perspective, the indigenous woman in her natural state was depicted once again as barbarous and unfit, unable or unwilling to raise her children to be good Christian citizens. In contrast, the writer argued, "The children distributed to families live happily, because their treatment and education in civilized practices . . . allow them to quickly mix with the other children."[106] The author's faith in the superiority of civilization and the weakness of indigenous family ties was so strong that he assumed that becoming civilized would provide more happiness than the mother-child relationship.

Despite that fact that nearly all Argentines at the time agreed that it was necessary to assimilate indigenous peoples, many reacted to the distribution system with disgust. *La Prensa* criticized the policy in general, arguing that the Indian, although an Argentine citizen, was being treated as if he were "outside the laws that regulate people's civil status."[107] A decade later, *La Nación* made similar arguments, claiming that indigenous peoples were being treated as slaves on sugar plantations in Tucumán.[108] A large portion of the negative reactions, however, objected to the distribution system over gender issues rather than a belief in indigenous peoples' rights as Argentine citizens. After the plan took effect, the Catholic paper *La América del Sud* claimed the separation of families violated the laws of nature, while Salesian missionaries reported fathers preferring to kill their children than be separated.[109] The chaplain of Martín García Island communicated the prisoners' sadness at seeing themselves separated from their children and recommended that this be avoided at all costs.[110] Daireaux's description above similarly implies the horror of such "heart-wrenching scenes," noting the children and mothers ripped apart and the obvious fright of the victims. Even *El Nacional* softened its position, describing the struggles of indigenous parents in detail: "terrified by that refined cruelty, that even their savage spirit cannot understand, they finally ceased asking those who would not be moved to have mercy."[111]

Even the most dedicated supporters of the destruction of indigenous cultures would eventually decry the breaking up of indigenous families. Within the Argentine Congress, which had previously encouraged such proposals, Representative Demaría argued in 1885 that the government had a moral obligation to stop the separation of families. In the same session, Representative Dávila proposed a declaration that "The House of Representatives of the Argentine Republic disapproves and condemns, as a system of placing Indians, this distribution of them that is done, separating mothers from their children, etc."[112] Mansilla and Zeballos were serving in the National Congress at the time, but at the moment neither took a firm stand against distribution. Mansilla admitted to being moved by the stories of misery but stated he would vote against the resolution because he believed it to be outside the bounds of Congress's scope of action.[113] Zeballos was not present for that day's session, but in 1888 he did publicly comment on the policy his research inspired when he lamented that distribution was "painfully shattering family ties, tearing children from their mothers' arms."[114] As a result of these complaints, the distribution system first passed to the control of the *Defensor de Pobres e Incapaces* (defender of the poor and incapable), where it was more closely regulated.[115] It was later phased out completely in favor of a system of colonies.

These critiques and the challenges they presented to the distribution system demonstrate the ways scientifically informed racial projects and gender ideology clashed. Zeballos, Daireaux, and the various newspapers unanimously disagreed with the distribution policy on gendered grounds, namely the cruelty of separating mothers and children. In contrast, they raised few complaints about the distribution of indigenous men and largely left unquestioned the state's right to dictate the occupation and living situation of its most marginal citizens. Daireaux continued to insist that indigenous people were the weaker race and that their destruction was natural and inevitable; Zeballos maintained that the Indians hated civilization and would only work under duress, even while he lambasted the steps taken to reach that goal.[116] Christian surety of the sacredness of the bond between mother and child collided with a scientific surety of the need to break that bond into order to put an end to barbarous cultures. Even though Christian-creole gender norms shaped anthropological science, in the end those same norms placed limits on the degree to which the policies it inspired could be carried out in practice. Ideas of family, men and women's roles in society, and appropriate sexuality were thus not only the root of Argentine scientific understandings of indigenous peoples and policies to control them, but also, in the end, a check on the viability of racially informed national projects.

✡✡✡

In their anthropological studies, the Argentine scientists highlight indigenous men's treatment of white and indigenous women, depicting their actions as barbarous and thus an impediment to their incorporation into the modern nation. Indigenous sexual behavior was particularly damning, precisely because it crossed racial boundaries. Tellingly, the anthropologists do not reflect critically on white men's treatment of women, which appears in their texts only as the imagined ideal against which they can measure indigenous men's failures. There is no acknowledgement of gendered problems in their own society, nor do they consider the fact that white men in Latin America had a long history of engaging in the same sexual behaviors they condemn in the indigenous men they study.[117]

Nonetheless, they themselves were white men who interacted with indigenous women on a near-daily basis as they traveled around the country. The next chapters will begin to untangle some of those relationships, including scientific interactions between the creole men and their female indigenous subjects as well the more scandalous realm of flirtation and desire. Because these elements are not foregrounded in the Argentines' texts, exploring them means skimming the margins of the written record and sifting through gossip, hints, and the unsaid.

CHAPTER 2

Sex and Specimen

Desiring Indigenous Bodies

PARADOXICALLY, DESIRE FOR INDIGENOUS bodies was inseparable from nineteenth-century Argentine efforts to craft a white, civilized identity for the nation.[1] In the second half of the century, fossilized indigenous skeletons grew in intellectual and commercial importance as scholars in Argentina and abroad insisted that the comparative study of ancient bones, particularly skulls, would provide a "satisfactory solution" to the debate over the birthplace(s) of the human race(s).[2] Scientists thus spoke at length of their desire to possess them. Many also were convinced that modern "savages"—that is to say, indigenous peoples—shared enough features with their prehistoric ancestors that their bones and skulls could stand in for those of their predecessors when necessary.[3] The confluence of past and present meant that, in addition to excavating ancient graves, River Plate scientists also raided contemporary indigenous cemeteries to satisfy their desire for possession. Even those they knew personally were not spared: in *Viaje a la Patagonia austral*, Francisco P. Moreno matter-of-factly recounts digging up the skeleton of his recently murdered Tehuelche "friend," Sam Slick, despite Slick's adamant refusal to allow Moreno to measure his head while alive.[4]

While the scientific desire to possess indigenous bodies has been studied extensively, less attention has been paid to the fact that it often coexisted with *sexual* desire. In *Una excursión a los indios ranqueles*, Lucio V. Mansilla wrote frequently of the attraction he felt for Ranquel women. His relationship with Carmen, one of his principal informants, is particularly marked by constant allusions to desire and its repression. Conversely, explorer-anthropologist Ramón Lista never expressed sexual desire in his works, but around 1890 he started a second family with a Tehuelche woman, engendering a daughter to whom he gave his name. This desire and its ramifications for the anthropologists' scientific-colonial projects are almost completely absent from the critical literature. Fernanda Peñaloza has demonstrated how Moreno's work explored the intimate lives of indigenous tribes and used their "sexual energy and erotic desire" to argue for the colonization of Patagonia, but neither she nor others have explored the sexuality of the anthropologists themselves.[5]

In this chapter, I draw on the work of scholars of colonialism, empire, and gender as well as the explicit testimony of the Argentine anthropologists to examine how sexual desire impacted the construction of scientific knowledge in Argentine anthropological texts. Common thought insists that the scientific process should be objective, but as far back as 1627 Francis Bacon linked eros and science. Far from advocating an indifferent objectivism, Pesic argues, Bacon insisted that "scientists must remain open to the heights of pleasure not only so that their sensibilities not become constricted but also for a strictly scientific reason. . . . In the alluring or the disgusting as well as in the indifferent may lie crucial discoveries which they cannot shrink from investigating."[6]

The Argentines linked their expeditions, actions, and conclusions to particular desires, some related to knowledge and others reflecting more bodily wants. I argue that Mansilla and Lista used eros as a potent lens for processing and understanding their relationships to a world that was simultaneously threatening, disappearing, attractive, barbaric, and in need of control. Mansilla's flirtations are principally narrative, deployed in a work that is very conscious of its constructed nature. In contrast, Lista's desire exists almost entirely off the page and can only be traced through gossip and oblique textual references. Lista's large oeuvre, however, permits us to compare his works before and after his affair to trace his changing attitudes. The differing conditions of their flirtations reveal the effects of intimate contact on representations of indigenous peoples and the creation of spaces for them within Argentine national identity.

"Methods of which we will not speak": Ramón Lista's Secret Life among the Tehuelches

Less well known than other nineteenth-century explorers and naturalists, Ramón Lista was a key figure in the development of Argentine science. He traversed Patagonia, Tierra del Fuego, Misiones, Chaco, and Salta and can thus be considered the late-nineteenth-century anthropologist with the broadest first-hand knowledge of Argentine indigenous peoples. The information he collected on these journeys resulted in dozens of publications, including monographs, official reports, articles in the popular press, and scientific pieces in domestic and international journals. This extensive production was cut short when he died under suspicious circumstances in November 1897 while exploring the Pilcomayo River in Salta province.

Lista was also intimately tied to the region's scientific and political institutions. He frequently spoke at the Argentine Scientific Society and cofounded the Argentine Geographic Society in 1881 with the explicit goal of encouraging exploration and disseminating scientific writing about the region.[7] Like all his contemporaries, Lista's scientific activities had political purposes. They supported Argentina's position in border disputes, planned colonization, and encouraged European immigration. Lista was even asked by the Departmento General de Inmigración (General Department of Immigration) to write a promotional report on the province of Santa Cruz. Published in Spanish, French, English, and German as part of the series Useful News for Immigrants, Workers, and Capitalists, Lista's document details the geography, fauna, climate, and development possibilities of the region.[8] In *Viaje al país de los onas*, he explicitly links these scientific activities to the creation of a white Argentina, writing, "the only motivation that guides me is that of opening new directions in scientific investigation, after which will come industry, commerce, and territorial expansion in latitudes where the Caucasian race that languishes and weakens under the burning sun of the tropics can easily establish itself and prosper."[9] Far from imaging a "pure" science, Lista was highly conscious of the political and racial applications of his work. Indeed, he continued to explore and study while serving as the second governor of the province of Santa Cruz from 1887 to 1892.

While governor and married to Agustina Andrade, a poet and the daughter of an influential journalist and politician, Lista entered into a relationship with a Tehuelche woman named Clorinda Coile. Andrade continued to live with their two children in Buenos Aires while Lista divided his time between the governor's house in Río Gallegos and Coile's toldo. Coile gave birth to a daughter in 1890 or 1891 and they named her Ramona Cecilia Lista.[10] Soon after Ramona's birth, Agustina committed suicide. Her death was not noted by any of the major newspapers of the era, despite her fame as a poet and the high standing of her family. In 1892, Lista resigned from his post and returned to Buenos Aires. The reasons for these actions are unclear. A century later, in 1981 the son of the interim governor who took Lista's place claimed that Lista resigned because he had become seriously ill.[11] Other accounts say that word of the affair had provoked scandal, and that Vice President Carlos Pellegrini sent a good friend of Lista's to Santa Cruz to convince him to resign and to drag him back to the capital.[12]

Either way, Lista's relationship with Coile was and has remained something of an open secret. In his 1898 travelogue, *La Australia argentina*, journalist Roberto Payró cites Lista's scientific work on the Tehuelches' creation myth, affirming that Lista "was very in contact with those Indians, so much so that he even lived among them, using methods of which we will not speak."[13] In a profile of Lista in the 1950s, Argentine historian Juan Hilarión Lenzi also reports the relationship, although he hedges it as gossip. The only thing he can say with certainty, Lenzi writes, is that Lista "abandoned" his post repeatedly to live in the toldos, "as a means of avoiding boredom and finding himself in the center of his investigations."[14] Both of these texts link Lista's scientific output to his romantic entanglements, yet critical works today continue to either ignore the relationship or only mention it as an afterthought.[15] Was the sexual relationship with Coile actually a "method," as Payró puts it, for gaining scientific knowledge? Did the experience of living intimately with the Tehuelches change Lista's anthropological understandings of the tribe and their role in the Argentine nation? Comparison of Lista's complete works from before and after the affair demonstrates a number of mechanisms via which sexual desire influenced his scientific methods and conclusions, shaping both the political projects he supported and his representations of indigenous peoples. Those images continue to circulate throughout the River Plate today.

Destined to Disappear: Lista's Assessments of Indigenous People before His Relationship with Clorinda Coile

Lista's first scientific work was published in 1878, beginning a prolific career that mixed writing for scientific journals with pieces for a more general audience published in the major Argentine newspapers of the day. By the time he was named governor of Santa Cruz in 1887, he had published twenty-eight texts detailing the physical and human landscapes of the Argentine interior. The period of governorship corresponds with a fallow period in his publication history, likely due to his administrative responsibilities and the time he spent in Coile's toldo. From his return to Buenos Aires until his death in 1897, he published thirteen additional works.

In the works from his first period of production, Lista offers a mixed assessment of the indigenous peoples he encounters. He represents some tribes as savage and difficult to civilize. In *Viaje al país de los onas* (1887), Lista lists the laziness, lack of religion, and possible cannibalism of the Onas (commonly called Fuegians during this period) of the southernmost Argentine territories as proof of the lowness of their race. Numerous European explorers had already arrived at the same conclusions. For example, Darwin's contact with the Onas led him to muse, "I could not have believed how wide was the difference between savage and civilized man: it is greater than between a wild and domesticated animal," and "Viewing such men, one can hardly make one's self believe that they are fellow-creatures, and inhabitants of the same world.[16] In his 1887 work, Lista writes that he agrees with Darwin's assessment, calling the inhabitants of Tierra del Fuego "a degraded race, that surely occupies the lowest level among all the savage peoples" (2:23).

In *Viaje al país de los onas*, Lista uses his observations to locate the Onas outside of present time. Although he describes interacting with them personally, he also emphasizes that they were "the faithful reflection of the primitive man of the quaternary caverns of Europe" (2:18). The insistence on the simultaneous present and prehistoric state of Argentine indigenous peoples is not unique to Lista's work, but rather a common rhetorical device found in texts by Moreno, Zeballos, Sarmiento, Florentino Ameghino, and others.[17] Lista extends this assessment to the Guaraní tribe living in the opposite extreme of the territory, using similar characteristics to determine that they, too, were "still submerged in the barbarism of the *Stone Age*."[18] In both cases, this "denial of coevalness," to use Johannes Fabian's term, excludes specific tribes from the nation even as Lista's physical explorations reaffirmed national control over their lands and paved the way for their appropriation.[19] Throughout his writing, Lista also explicitly contrasts this primitive nature with the "intellectual superiority" of the "European element," laying the groundwork for the colonization of southern Patagonia and further transformation of the region's ethnic, economic, and political profile.[20]

In contrast to his negative representations of the Onas and Guaraníes, Lista was relatively impressed by the Tehuelche Indians of Northern Patagonia. Like his peers, he believed they were a more civilized race that had displaced the Onas from Patagonia in prehistoric times, pushing the latter south (2:23). In the report from his 1879 expedition, he describes them as docile, respectful, hospitable, and not at all a danger to creole Argentines.[21] In fact, like Estanislao Zeballos stated one year prior, Lista argues that the Tehuelches could be settled in colonies and taught to be good farmhands and gauchos.[22] Despite this positive representation, Lista also expresses great sympathy for the Argentine and Chilean merchants who had to spend time in Tehuelche camps. He describes living there as a "sacrifice" due to the filth of Tehuelche toldos, violent demonstrations of affection, and the constant sounds of old women singing and dogs fighting.[23] From these descriptions, it is clear that Lista's acknowledgement of the future potential of Tehuelche peoples did not imply their current membership in the national body. In order to be incorporated, they would have to change significantly.

In his early works, Lista further distances all indigenous tribes, including the positively valued Tehuelches, from the Argentine nation by contending that, "obeying a fatal law, [they] march rapidly toward their extinction."[24] This rhetoric, common to many River Plate ethnographies of the time, cast the disappearance of indigenous cultures as a natural consequence of evolution. They could not play a role in the modern nation, for they were bound to disappear. Progress would move Argentina and some of its inhabitants forward, but the indigenous line would be truncated.

Invoking evolution allows Lista to present the disappearance of indigenous peoples as inevitable, but also natural and thus amoral. In this way, he is able to excuse the ways in which his own actions hastened the vanishing of the Argentine Indian and his and her culture. In several instances during his explorations of Tierra del Fuego recounted in *Viaje al país de los onas*, Lista describes taking indigenous women and children prisoner and making them act as interpreters and guides (2:63, 81). "In order to distinguish between them," he gave them Christian names that would later be confirmed by baptism (2:63). Through these actions, Lista slowly erased the Ona's indigeneity, a micro-experiment proving his assertion that they might be able to become "useful men" (and women!) if they were removed from their huts and all their "barbarous customs" were replaced with civilized ways (2:48). Lista also participated in the violent elimination of the Onas: in November 1886 he authorized his men to open fire on a group, resulting in the death of twenty-eight (2:60). Despite censure from the missionaries with him, this episode did not particularly change his behavior during the rest of the expedition. When writing about these incidents after the fact, Lista was able to justify his actions and escape accusations of excessive cruelty precisely by presenting indigenous extinction as predetermined by natural law.

Unnatural Extinction: Lista's Post-affair Defense of Indigenous People

These attitudes and actions are an unlikely prequel to a long-term sexual relationship with an indigenous woman, yet that is exactly what developed between Lista and Clorinda Coile. Lasting a number of years and resulting in a child, this intimacy had a profound effect on his scientific representations of Coile's tribe, the Tehuelches. While his earlier texts highlighted indigenous inferiority, located them outside of the nation, and claimed biological factors doomed them to extinction, his first text after the affair, *Los indios tehuelches: Una raza que desaparece* (1894), is a powerful defense of the Tehuelches' right to live and an emotional condemnation of their treatment by Argentines and Chileans alike.

He begins the text by explicitly distancing himself from his previous statements, establishing an apologetic tone. The opening paragraphs present nineteenth-century Argentine nation building as the latest iteration of violent Spanish-indigenous relations in the Americas, paralleling the great cruelty of the sixteenth-century conquistadores with the "modern conquest of the Pampa."[25] Lista develops a long list of past and present abuses and juxtaposes it with conquest-era figures who distanced themselves from those horrible actions: Pedro de la Gasca, Bartolomé de las Casas, Juan Polo de Ondegardo y Zárate, and the Bishop Vicente de Valverde y Alvarez de Toledo. He then presents himself as the modern progeny of those defenders of indigenous peoples, a lone voice in the wilderness whose dissent would later leave him on the right side of history. He concludes with a fervent desire that his humanitarian sentiments would be well received in Chile and in Argentina (2:129).

Lista uses a number of strategies in the text to prove the Tehuelches' humanity, criticize Argentine and Chilean actions in Patagonia, and call for new policies defending the Tehuelches' right to live. One of the most important is a revised ethnography of the tribe that contradicts both his contemporaries' and his own prior assessments. Lista puts Tehuelche oral tradition down on paper, including a long recounting of their origin myth and several other legends (2:131, 144–45). Whereas he had previously claimed that there was no Tehuelche religion worth recording, he now makes careful note of their god-like figures, superstitions, burial habits, and understandings of the afterlife.[26] Although he maintains that their religious understandings are basic, he now knows much more and ascribes value to this knowledge. He similarly records Tehuelche songs, dances, clothing, and language, including a dictionary that is much larger than those in his previous works and includes new phrases such as "This is my child," "This is my wife," and "I love you" (*Los indios tehuelches*, 2:169–81).

This new information and more positive interpretation of Tehuelche life is closely connected to Lista's sexual relationship with Clorinda Coile. The relationship took place during several years among the toldos, far surpassing the several months he drew upon when writing *Viaje al país de los tehuelches* in 1879, or Moreno's similar experience in the mid-1870s. The greater amount of time spent among the Tehuelches facilitated his more in-depth observations of their culture, and it is likely that Coile herself served as a principal informant. Indeed, both George Chaworth Musters and Lucio V. Mansilla had previously noted that indigenous women were their primary sources of information about the languages and/or cultures of the communities they studied.[27] And although we cannot know for sure, we have to wonder if Lista would have included intimate phrases such as "this is my child" and "this is my wife" had he himself not been part of Tehuelche family life.[28]

Lista himself argues for the importance of these connections. In the "Two Words" that open *Los indios tehuelches*, Lista notes, "The following pages have been written under the hut of the Patagonian savage, always hospitable and caring with the traveler. To fully understand the life of the tribe, to sound out the thoughts and heart of the children of the desert, I have had to overcome many obstacles, stoically suffering the slander and perfidy brought by the two great virtues of these decadent times: envy and malice" (2:125). "Slander and perfidy" refer to gossip about his relationship with Coile, and Lista's critical attitude makes clear that he views his extensive time with the Tehuelches as a boon to his science. Personal ties do not harm science by violating objectivity, he insists, but rather provide the opportunity for the kind of extensive study that leads to truth (2:125).

At least some of his contemporaries appear to have shared this belief. A few years later, Payró judged that Lista's work on the Tehuelches was his most valuable and most trustworthy, "due to the methods he used in order to gain access to the customs and the intimacy of those Indians."[29] Although there is a note of sarcasm in Payró's observation, his use of the explorer's work in *La Australia argentina* suggests that he did truly believe that Lista was more capable precisely because of his near-insider status. The "methods of which we will not speak," to use Payró's phrase, paid scientific dividends.

At some moment between 1887 and 1894, a period that corresponded to his life in the toldos as a lover and father of Tehuelches, Lista also reassessed the factors contributing to their disappearance. In *Los indios tehuelches,* he writes that he now believes that they are not dying because of

> the natural law of evolution, but rather due to gunpowder and liquor, the unstoppable cruelty of some and the reigning predation of others. What is happening is truly inconceivable; one could say that a divine curse weighs on them: they are the original owners of the lands they inhabit and this land does not belong to them, they do not even possess a parcel where they can rest at the end of the day: they have been born free and are now slaves; yes-

> terday they were robust and of large body: today tuberculosis kills them and their stature shrinks. Everything is against them, the abyss surrounds them, they are going to disappear. (2:128)

He explicitly rejects a biological cause for this disappearance. While extinction is a natural consequence of evolution, Lista argues that "extinction is recasting, incorporation, not unstoppable and cunning annihilation by some instinct of civilized evil, tacitly permitted by those who govern" (2:127). In other words, Lista defends the idea of extinction but claims that anyone that attributes the disappearance of Tehuelches to natural evolution is intentionally misunderstanding the term. The Argentines' and Chileans' actions were not justified within the struggle for life, but were excessive, immoral, and inherently unnatural.[30]

Throughout the text, Lista explicitly points out and shames the actors he sees participating in this great cruelty. Whereas previously he sympathized with the creole merchants who suffered the filth of the toldos, in *Los indios tehuelches* he identifies with the Tehuelches and depicts the merchants as disgusting barbarians. He accuses them of raping women, perverting children's morality, robbing the Tehuelches, and teaching them all of the bad aspects of civilization and none of the good (2:128). He then inverts the civilization-barbarism dichotomy and argues that despite their use of Spanish and fancy clothes, the merchants "in reality are more savage than the Indians, being as they are their corruptors and plunderers, with no brake whatsoever to slow their attacks and theft, and without any law that punishes their constant crimes." In fact, he explicitly identifies the merchants of Punta Arenas and Río Gallegos as "the two elements of extinction," showing once again the unnaturalness and cruelty of the disappearance of the Tehuelches (2:128–29). The Argentine and Chilean governments are complicit in this tragedy, for they have turned a blind eye to the abuse, "[watching] them die with the same impassiveness as Caesar watched the gladiators die in the circus. . . . 'Those who are about to die salute you,' the Tehuelches could say to the Argentine legislator, and also to the Chilean one" (2:128).

Science and personal relationships also structure Lista's suggested response to the crisis. On one hand, Lista continues to view time and development as unstoppably forward-moving and resigns himself to the disappearance of the Tehuelches. He embraces the common trope of the Vanishing Indian, speaking of "the relics of Tehuelche race" (2:129). "Today," he laments, "everything has ended or is going to end for the Tehuelche: the shepherd repels him, the sheep grazes where the guanaco used to do so. Everything is against him: the governments abandon him, and the Christian settler, ruthless, gets him drunk in order to strip him of all he possesses. Terrible destiny!" (2:165). In this passage, he foresees the end of the Tehuelche as a culture, for their way of life, possessions, and land were being taken away by the colonization he himself had previously encouraged.

But *Los indios tehuelches* is not just a eulogy. It is also a call to action. Contrary to the passive cruelty of the government, Lista pleads for action to physically save some Tehuelches. He couches this need in both emotion and logic, arguing that if the Argentines did not understand the humanitarian reasons to protect the Tehuelches, then surely they could appreciate the scientific interest in doing so (2:128). The solution, he claims, was to follow the example of the United States and create indigenous reserves where they could be protected:

> Both countries [Argentina and Chile] should enact an agricultural reserve law, modeled on the text of the North American law in favor of the Sioux, prohibit under severe punishment the sale of alcohol in the Indian encampments, create children's schools in the same camps under the direction of virtuous missionaries, and both governments will not have anything but reasons to rejoice, if there can be any rejoicing in a noble action, in extending a hand to one who is already on the edge of the bottomless abyss. (2:129)

Isolating them into reserves and prohibiting the sale of liquor could cut down on violence and other crimes, while primary schools run by missionaries could focus on bringing the positive elements of civilization to those groups.

Lista's suggestions make clear that he was concerned primarily with the physical preservation of the Tehuelches and significantly less worried about the longevity of their cultural habits. The steps he proposes to help the Tehuelches would erase many elements of their way of life, including nomadism and their religious beliefs, replacing them with more typically Argentine habits. His efforts were motivated not by a certainty that the Tehuelches in their current state were equal to the Argentines, but rather by the conviction that they were at least human and thus deserving of some compassion. Lista's plan called for the elimination of those "unnatural" aspects of extinction he had railed against previously, leaving space for creole culture to absorb the Tehuelches and lead to a natural extinction.

In *Los indios tehuelches,* Lista also develops a positive vision of *mestizaje* (racial mixture) that directly challenges the fear seen in Zeballos's works or in that of the European scholars who predicted the inevitable downfall of the Hispanic hybrid. In an earlier text, Lista noted that in Viedma, the "well-known intellectual superiority" of the Europeans guaranteed the good results of mixing between Argentines and Europeans, and that those children were generally more beautiful, more virile, and smarter than the children without this mixed racial background.[31] In *Los indios tehuelches*, Lista intensifies his calls for mixture and presents it as a way to help the Tehuelches, even while slowly eliminating their culture. He observes that in contrast to plummeting Tehuelche birth rates, "the union between an Indian woman and a white man is more prolific" (2:150). This simple statement shows that creole-indigenous relationships were eugenistic, proving the Tehuelches' humanity, and contradicts negative assessments of the Hispanic hybrid. Again, his personal experiences—born of sex—bore this out: Lista and Coile had produced a child who Lista gives no reason to suppose was in any way inferior.

He further argues that the Argentines and Chileans should allow the Patagonians to slowly "melt into the civilized masses" through racial mixing" (2:128). While this mixture would still lead to the eventual disappearance of the Tehuelches as a separate race or culture, he insists that this incorporation or recasting ("refundición") was analogous to the workings of evolution and extinction in other biological realms. It was thus the natural and therefore moral choice. It is clear that the mechanisms for this *refundición* included the very activity that most terrified many of Lista's peers: interracial sex. In order to be incorporated, indigenous men and women would have to have heterosexual relations with their creole counterparts. Children born of these unions would not be accidental or problematic, but rather a path to the future.

To make these unions acceptable, Lista inverts the racial identity of the lovers: whereas earlier and contemporary texts feared the rape of white women by indigenous men, Lista explicitly advocates for pairing Tehuelche women and white men (2:150). Lista's inversion reflects scientific, cultural, and political ideas of his time. Racial scientists worldwide were curious about the relative fecundity of crossings between men of inferior races and white women and vice versa.[32] Parallel efforts to understand the mechanisms of heredity also were deeply invested in gender. Early theories attempting to explain the similarities and differences between generations assigned men and women active and passive roles, respectively.[33] Although by the late nineteenth century many of these ideas had been replaced by theories that acknowledged the equal importance of maternal and paternal inputs, remnants of older ideas still circulated. In this framework, a relationship between a white Argentine man and an indigenous woman was relatively harmless for she would be little more than the nurturing environment in which the male's traits would be reproduced. Indigenous women would thus impart little to the future generation. Indigenous men's rape of white women was so problematic for this same reason.

More contemporaneous explanations of heredity recognized women's active role, but, like Lista, emphasized the importance of the gender of each individual to mixed-race crossings. For example, the German scientist Ernst Haeckel, who was very influential in Argentina, argued that "the hybrid produced by the crossing of two different species is a mixed form, which receives qualities from both parents; but the qualities of the hybrid are different, according to the form of the crossing. In like manner, mulattoes produced by a European and a negress show a different mixture of characters from the hybrids produced by a negro with a European female." While presenting these observations as fact, he did not name specific, consistent differences nor was he able to explain why these differences occurred.[34]

Lista and his Argentine compatriots similarly insisted on the gendered aspect of hybridity, but imbued crossings with specific values in line with other, extra-scientific factors. Lista's prescription was likely an extension of his own personal experience: he was a white man who had a child with a Tehuelche woman. It also echoes broader patriarchal and racial biases of the era. Most Argentine male elites' work demonstrates their confidence in both men's power over women and the inferiority of indigenous peoples. The interplay of these two convictions meant that for Lista and others who supported interracial sex, white men's unions with indigenous women could paternalistically help the latter rise to civilized levels. Conversely, indigenous men could only degrade white women. Such unions could not improve the nation, and thus were to be feared and avoided.

Lista's specific recommendations also had important political ramifications. By specifying the gender/racial composition of interracial sex, Lista ensured that white men like himself stayed in control of the process by preventing changing attitudes from fundamentally shifting power structures. Encouraging indigenous men and white women to reproduce meant granting the power to decide to one or both of those "inferior" groups, thus acknowledging their independence and agency. Given that both women and indigenous men were often associated with irrationality, it would also mean turning over an important element of nation building to individuals who could not be entrusted to exercise caution. Consequently, even as Lista advocated for interracial sex as the most important driver of evolution and a potential resolution to Argentina's racial growing pains, relationships between creole women and indigenous men continued to be unthinkable, at least for porteño elites.

Sex and Repression in Lucio V. Mansilla's *Una excursión a los indios ranqueles*

Originally published as a series of letters directed to a Chilean friend, Santiago Arcos, and printed in the national newspaper *La Tribuna*, Mansilla's *Una excursión a los indios ranqueles* appeared as a stand-alone book at the end of 1870. It was immensely popular at the time of its publication. Since then, it has never been out of print.[35] In *Una excursión*, Mansilla relates the details of an eighteen-day expedition he led from Río Cuarto in the province of Córdoba to the encampments of the Ranquel cacique Mariano Rosas in Leubucó. Mansilla states that he undertook the journey in order to ratify a peace treaty with the Ranqueles; Silvia Fernández has shown he was also avoiding court martial for the improper execution of a deserter.[36] In the end, Mansilla convinces the Ranqueles to sign the treaty that gives pretext to his excursion and leaves professing a new respect for the "barbarians" of the pampas. He insists that although the Ranqueles must be subdued in order to better the nation, they should not be exterminated.

A multi-genred text, *Una excursión* has been approached by scholars as creative non-fiction, memoir, detective tale, and military-political treatise. It is also, as Carlos Alonso, Andrew Brown, Homero Guglielmini, and others have suggested, an ethnography, or at least it aims to be one, relying heavily on phrenology and other human sciences.[37] Mansilla grounds his narrative within a global scientific discourse, especially with regard to questions of race, ethnicity, and human development. He collects data regarding the Ranqueles' language, marriage customs, governmental structures, ceremonies, and more. For this focus, Gillies calls him an ethnographic precursor: "Mansilla was writing a few years before Morgan, Tylor, Frazer, and the other founding fathers of modern anthropology, yet his voracious intellectual curiosity led him to ask an ethnographer's questions."[38] Mansilla's contemporaries also acknowledged the scientific value of the work, and *Una excursión* won a prize at the International Geographic Congress of Paris in 1875.[39]

Whereas Lista's encounters with interracial desire took place outside of his narratives, Mansilla's frontier flirtations are principally textual. In *Una excursión*, Mansilla embodies Bacon's eros-science approach in the nineteenth century. He does not depict his scientific and political goals as separate from bodies, desire, and pleasure, but rather delves deeply and personally into the disgusting and the appealing. In fact, Mansilla presents his excursion as two competing conquests: his attempts to win over the Ranqueles and barbarism's efforts to win him over. These endeavors are described as seductions and narrated in sexualized language that involves both metaphorical gender relations and actual physical relations inscribed on indigenous women's bodies. These manifestations of desire are not distinct from the anthropological sections of his text but rather are absolutely integral to Mansilla's scientific study. The ways in which Mansilla narrates these two competing conquests and his own feelings of desire influence his pronouncements with regard to the Ranqueles' place in the human family as well as their positioning in his stadialist understanding of human development. Nonetheless, the highly constructed nature of Mansilla's text and his own later actions suggest that in contrast to Lista, Mansilla's representation of sexual desire and the concomitant change in attitude is largely rhetorical and thus only temporary.

The reader of *Una excursión*, primed by Sarmiento's *Facundo* and other colonial-scientific texts of the era, expects to encounter a narration of how civilization seduces barbarism. Indeed, Mansilla sets out to win over the Ranqueles. He situates his journey within the common trope of the feminized land waiting for the masculine explorer-colonist to subdue and then fertilize it. This tradition, which Anne McClintock calls an "erotics of ravishment," dates back to at least the fifteenth century.[40] In *Una excursión*, Mansilla relies on this well-worn narrative technique to narrate his journey across the pampas. He depicts the land as feminine and positions himself as the physical representation of masculine power and technology come to conquer it.

In many Argentine texts it is empty space that is depicted as feminine through its virginity or barrenness, but Mansilla also feminizes the very heart of Ranquel civilization itself. His journey from Río Cuarto to Leubucó is described as a journey to a highly desirable center. As Mansilla approaches the toldos, he writes, "I shall at last penetrate into the forbidden enclosure" ("Voy a penetrar, al fin, en el recinto vedado"; *A Visit*, 105). The description of the Ranquel encampment as a forbidden enclosure to be penetrated by Mansilla has clear sexual overtones. The Ranquel toldo becomes a metaphorical vagina in a feminization further underscored by Mansilla's equating of women and the Ranquel Indians through their association with secretiveness and the unknown.[41] In his masculine civilizing role, he aims to seduce the feminized Mariano Rosas and the Ranqueles into revealing their secrets, which he would then make explicitly visible in a variety of formats associated with masculinity: military memos, maps, the travel narrative itself, and scientific theorizing. In this way, Mansilla's depiction of his excursion enacts the European trope of the discovery of knowledge as "the male penetration and exposure of a veiled, female interior; and the aggressive conversion of its 'secrets' into a visible, male science of the surface."[42]

Throughout the narrative, Mansilla chronicles many of the moments when he—and the culture with which he identifies—successfully seduce the Ranqueles. Upon arriving at Leubucó, he is amused to discover that Mariano Rosas is a regular reader of *La Tribuna*, the very newspaper in which he will later publish his narrative (*A Visit*, 48, 218). Similarly, some Ranqueles have excellent table manners and refuse to eat without spoons, napkins, and the other accoutrements of civilized dining. In these moments, Mansilla sees civilization already taking hold in the tribe and remarks, "Positively, it's not so difficult to civilize these barbarians" (99). Mansilla himself has at least three moments of successful conquest. First, he is permitted to enter Leubucó, the so-called forbidden enclosure. Mansilla attributes this achievement to his ability to get the Ranqueles to see him as an equal and thus trust him, seducing them with liquor, food, and displays of masculine strength. Second, after nine hours of speaking to the tribe about the benefits of the peace treaty and the good intentions of the Argentine government, Mansilla convinces Mariano Rosas to accept the agreement (304). Finally, after significant back and forth with the chieftain, he arranges the release of Macias, a primary school friend who had been taken captive by Rosas in 1867 (349). Mansilla views each of these moments metonymically, equating his own accomplishments with the victory of civilization over barbarism.

While these attitudes are expected, Mansilla also departs from this norm by describing some of his interactions with the Ranqueles as the seduction of the civilized male by feminized barbarism. As he journeys across the pampas, he revalues frontier life and takes a hard look at his own culture. While acknowledging that "civilization has, no doubt, its advantages over barbarism," he recognizes that these are "not so many as those who call themselves civilized assert." Civilization means many doctors and lawyers, but also many ill people and lawsuits. While it is certainly "of all modern inventions, one of the most useful for the well-being and progress of man," it is plagued with problems that are overlooked in the push toward ever-higher levels of culture (45-46).

In contrast, Mansilla finds much to appreciate in Ranquel society. He is impressed by their system of *vuelta*, whereby an individual receives whatever they need from the community under the assumption that they will pay the debt back when they are able. Mansilla remarks ironically that the Indians have established the Golden Rule, while among Christians "a hungry man does not eat if he does not have the wherewithal" (271-72). He also admires their humane way of killing animals and other characteristics that frequently evoke Jean-Jacques Rousseau's concept of the noble savage (185).[43] The combination of Mansilla's critiques of the trappings of civilization and appreciation for barbarism represent the latter as a rejection of the challenges, hypocrisies, and stifling norms of nineteenth-century creole Argentine society and a return to a simpler time. With this inversion, Mansilla questions the overwhelmingly optimistic view of progress embedded in stadialism and positivism. Interestingly, Lista expressed a similar preference for the barbarous countryside. In his works, he juxtaposes the peace of the frontier with the complications of Buenos Aires and expresses a desire to live like a savage: "I would like to still live a long time as a nomad, go to sleep wrapped up in my fur cape, climb the tall hills and jump the waterfalls."[44] As Mansilla pronounces, it was easy for a civilized man to give in to the temptations of the wild: "By living among savages I have come to understand why it has always been easier to pass from civilization to barbarism than from barbarism to civilization. We are very arrogant. And yet, it is easier to turn Orión or Carlos Keen into an Indian chief than to turn Calfucurá or Mariano Rosas into an Orión or a Carlos Keen" (*A Visit*, 70–71).[45]

Mansilla's experiences on the frontier show this transformation in action: Doña Fermina, captive for decades, confesses to him, "I look like a Christian, because Ramón allows me to dress as they do; but I live like an Indian woman and, honestly, I think I'm more Indian than Christian" (360). Once a part of creole society, Doña Fermina had been transformed by her contact with the Ranqueles and the desert. Similarly, many renegades and gauchos felt loyalty toward the Indians and had adopted their habits, even taking Indian wives. One man is an Argentine "married among the Indians, whose habits and customs he has adopted to such a degree that one might doubt he is a Christian. . . . José is tied to the wilderness by love, family, and wealth" (209). Similarly, the mestizo Mora "knows that civilized life is better than life in the wilderness; but, as I have said already, he is bound to the wilderness till his dying day by love, family, and property" (230).

Mansilla's descriptions of these men and their life stories reveal a key aspect of *Una excursión*: The seductions of barbarism are not just about freedom. They are also sexual. Mansilla's favorable impressions of these men and long digressions narrating their lives normalize interracial relations, so often elided in Argentine works. He leaves no room for the reader to doubt the nature of these ties when he recounts a conversation with Miguelito in which the gaucho expounds on the sexual freedom of indigenous women. Miguelito concludes, "Well, then—why do you think Christians like this land so much? There has to be a reason!" Mansilla does not condemn Miguelito's statement on sexual desire but actually seems to ratify it by musing, "He left me thinking about the enticements of barbarism" (147-48). Ironically, the interracial attraction that Zeballos and Mansilla himself use to condemn indigenous men as too masculine and thus barbarous becomes a tool for erasing distinctions when the genders of each race are inverted. The possibility that nonindigenous men's relationships with indigenous women might be coercive never once appears in Mansilla's text, despite ample examples of the reverse.

Further paving the way for the seduction of white or mestizo men by indigenous women, Mansilla depicts the frontier as a sexually liberal space defined at least partially by female agency and pleasure. He describes Ranquel marriage customs in detail and stresses the sexual freedom of single indigenous women. As he describes it, a Ranquel woman gives herself "to whomever she likes best" and "neither her father, her mother, nor her brothers will say a single word to her about it. It is no business of theirs, but hers alone. She is mistress of her own will, and of her body: she can do with it what she likes. If she yields, she is not dishonored, or criticized, or looked at askance. On the contrary, it is a proof of her value." Despite being contrary to creole norms, Mansilla makes Ranquel sexual practices acceptable by presenting this sexual freedom as natural fact. Carmen tells him all about her "adventures with a certain youth" and Mansilla editorializes that "humankind is subject to the law of reproduction. There is nothing strange in the fact that she, being a free woman, should give herself to whom she pleased and that from one day to the next she should find herself a mother" (224). Like evolution and other natural laws, casual relations are a result of biological imperative and thus should not be viewed as moral failure or evidence of the Ranqueles' inferiority. Importantly, however, like Lista Mansilla restricts his positive musings on interracial sex to pairings composed of white men with indigenous women; white women connected to indigenous men in the toldos continued to be represented as victims of an irrational and primitive sexuality.

Christina Iglesia describes Mansilla's preference, as a civilized man, for barbarism as scandalous; even more shocking is that he was tempted to participate in these frontier relations despite presenting himself as a representative of upper-class, European, civilized Argentina.[46] Whereas Lista generally described indigenous women from a distance, Mansilla narrates several personal interactions with specific women. He repeatedly describes them as beautiful in a tone that is more sexual than aesthetic. In one passage, he describes two of Carmen's sisters in the following fashion:

> The two women looked magnificent: their skin shone like polished bronze; their long braids, black as ebony and adorned with ribbons of local weave, fell gracefully over their backs; their short, even, naturally clean teeth gleamed like ivory; their little hands had short, shapely, tapered fingers; their little

> feet, the nails well pared, were perfectly clean. That morning, as soon as the sun rose, they had gone to the edge of the lagoon, taken a short bath, and, withdrawing a little from us, painted their cheeks and lower lips with carmine, which the Chileans bring them and sell them at an extravagant price. María, Villarreal's sister-in-law, more concerned with her appearance than her married sister, had given herself a few little beauty spots, a very favorite adornment with Indian women. (99)

The first two aspects that Mansilla lists as evidence of their beauty are those that most obviously characterize the women as indigenous: their darker skin and hair, the latter of which is archetypically styled in braids. Their beauty is a result of their racial background, not despite it. Furthermore, he directly contradicts the stereotype that indigenous tribes were filthy, casting aside an impediment to desire. The women's teeth are so white and shiny as to look like ivory, their toenails are carefully cut, and their feet perfectly clean. He explicitly mentions the women taking a bath in the lagoon. In additional to highlighting their hygiene, this brief allusion to a moment of nudity invites the reader to imagine the scene, thus participating in the frisson of desire the women provoke in Mansilla.

Other passages are even more explicitly sexual. He describes Villarreal's sister-in-law as a very pretty woman who dresses less modestly than her sister. When she lifts her arms, he writes, "one could see the hair in the hollow of her armpit, and the folds of her chemise, widening, revealed part of her bosom" (96). Whereas Lista matter-of-factly described young indigenous women's attractiveness, Mansilla inscribes his positive assessment of Ranquel beauty in a context marked by flirtation, pleasure, and sex.

He also presents his own reactions as typical of his entire culture. It is not his personal idiosyncrasies that drive his admiration of Ranquel women; other civilized men would clearly see their appeal. The cacique Ramón's daughters "were at the peak of their physical development, and a man of taste anywhere would have looked at them a long while with pleasure" (362). Similarly, Villarreal's sister-in-law "would be a brunette to make quite a few hearts ache if she put on creole clothes" (96). These generalizations add legitimacy to Mansilla's seduction by barbarism and force the reader to validate his experiences of frontier desire. Finally, after describing the Villarreal sisters, Mansilla makes clear that the Franciscan missionaries slept next to his bed. He does so "to forestall the reader's malicious thoughts," once again belying the myth propagated by Zeballos that racial mixing was largely the result of forced captivity or the province of the lower classes (97). Mansilla thus paradoxically affirms interracial sexual possibilities even while denying than anything happened in this particular instance.

Mansilla's Scientific-Sexual Thesis

The enumeration of the seductions of barbarism and barbarous women is offered as proof of Mansilla's scientific theses. Early in *Una excursión*, he makes clear that he means to proffer opinions on the debate about the unity of the human species and the destiny of the different races. To do so, he engages Santiago Arcos as narratee. Born in Chile in 1822, Arcos had traveled widely through Argentina and published two texts pertaining to the Argentine Indians: "Cuestión de indios," a short essay from 1860, and the more extensive *La Plata: Étude historique*, published in Paris in 1865.[47] In both, he advocated for an offensive war against the Indians designed to eliminate them and thus end the threat to life and property on the frontier. In the first chapters of *Una excursión*, Mansilla presents his scientific thesis by invoking prior personal conversations with Arcos. He indicates that he had changed his mind regarding the unity of the human species and the "historic destiny of different peoples." He goes on to explain that while he once believed that civilization was meant to battle constantly to preserve itself against barbarous human races, now he affirms, "I believe in the unity of mankind" ("Creo en la unidad de la especie humana"), placing him in opposition to Arcos (*A Visit*, 12).

This statement is repeated throughout the text in various forms, often using the metaphor of humankind as a tree with one trunk or as a plant that is a single species (366). By asserting that all the human races formed one species, Mansilla also affirms a single origin and the possibility of reproduction between the different races. This assertion of the single origin of humanity (monogenism) went against the polygenism espoused by luminaries such as Georges Cuvier, Louis Agassiz, Samuel Morton, and Robert Knox. In fact, in 1870 Mansilla's assertion of monogenism would likely have been the minority opinion among the European and North American scientists engaging in early studies of physical anthropology.[48] Instead, Mansilla follows in the tradition of Alcides d'Orbigny by asserting a common origin for all of mankind, including the seemingly barbarous tribes of the Americas.

Mansilla's descriptions of indigenous women and sex on the frontier support the assertion that the different races were varieties of a single species rather than distinct species in three ways. First, he establishes that women's actions, particularly their relationships to men, are the same everywhere. Two young indigenous women "flirted just as two Christian girls would have done" (362). Others approach Mansilla without fear, saying that since they were women, they had no reason to be afraid. This statement leads Mansilla to bemoan, "Lord, these women—they feel safe everywhere! And meanwhile, where are they not at risk?" (71). Overall, he concludes that "women are the same under every constellation . . . whether on the banks of the majestic Río de la Plata, or on the wide plains of the Pampas" (332–33). Like James Cowles Prichard (whose work he owned), Mansilla viewed this uniformity of behavior as evidence of the unity of the species; sexual relations are the field that make it obvious.[49]

According to Mansilla, philological study further confirms his conclusion. He notes linguistic similarities between French and Araucanian and muses, "French philologists are at liberty to investigate whether these words were taken by the Indians from the Gallic tongue, or the other way around! I can only say it is very curious that, among both Indians and Frenchmen, the words *cancanear* and *cancan* are related to Cupid and his temptations" (194). Although he refuses to say so definitively, Mansilla's philological reflections suggest that sexual mores are the same in both France and the Argentine interior, equating one of the most respected cultural touchstones for elite Argentines and the "barbarous" Indians in affairs of the heart. If women's beauty, actions, and beliefs are the same in Leubucó and Buenos Aires (and even in France), what evidence can there be that they belong to different species?

Mansilla further argues for the unity of the human species by juxtaposing the huge gender gap he perceives with minimal racial differences. He depicts women and men as continually at odds, with women everywhere using their feminine wiles to dominate men (6). He sympathizes with the mixed-race gauchos on the frontier precisely through stories of their troubles with women. Within the toldos, indigenous women's actions also suggest the need for solidarity between Mansilla and indigenous men. Over and over, he contrasts unfaithful, unknowable women in possession of a "diabolic power" with a masculine "us" suffering at their hands across racial and class lines.[50] By claiming that all women are troublesome, he unites men such that, as María Rosa Lojo claims, the male Indian is "substantially identifiable with any white male, and woman, without denying her belonging to the human species endowed with a soul, is the *other*, a distinct and always somewhat distant being (although she reaches points of maximum physical and affective closeness)."[51] This masculine solidarity in opposition to women is so strong that several critics have argued that homoeroticism is one of the structuring themes of the text.[52]

Finally, Mansilla turns to hybridity to establish the unity of the human species. In the late nineteenth century, one of the most frequent characteristics used to determine if two groups were varieties of the same species or different species altogether was their fecundity. French anthropologist Paul Broca, for example, argued in *On the Phenomena of Hybridity in the Genus Homo*, that "in the genus homo, as in the genera of their mammalia, there are different degrees of homoeogenesis, according to the race or species."[53] These degrees ranged from agenesic crossings, or "Mongrels of the first generation, entirely unfertile, either between each other, or with the two parent species," to eugenesic crossings, or "Mongrels of the first generation entirely fertile" both with each other and with their parent species.[54] When Mansilla notes the attractiveness of indigenous women and the number of interracial couples on the frontier, he demonstrates that desire and sex absolutely cross racial lines. These unions are eugenic, as demonstrated by the huge number of mestizos and children of mestizos that Mansilla meets during his excursion, and thus are further evidence that indigenous peoples, creole Argentines, and mestizos are all part of one human species.

In addition to proving the humanity of the Ranqueles, Mansilla agrees with Lista that sex and the resulting fusion of races is the key to the problems of the frontier. Throughout *Una excursión*, he criticizes those who advocate for the extermination of the Ranqueles. He points out the hypocrisy of those who study in order to "despise a poor Indian, to call him a barbarian, a savage; to ask for his extermination, because his blood, his race, his instincts are not easily assimilated to our empirical civilization—a civilization that calls itself humanitarian, upright, and fair-dealing, though those that kill by the sword die by the sword; that bloodies itself on grounds of self-esteem, of avarice, of pride" (366). He also argues that there was no scientific basis for these assertions, for the stakes were too high to rely on scanty information about racial characteristics. Instead, he suggests that the Ranqueles should be incorporated, for "Today it seems to be a point beyond dispute, that the fusion of races improves the condition of mankind." And the Ranqueles, he finds, are exceptionally apt for this incorporation, because they are "a solid, healthy, well-constituted race, without that 'Semitic' persistence that wards off all other races from any tendency to intermingling and crossbreeding" (383–84). Contrary to those who feared the effects of miscegenation, Mansilla advocated encouraging desire in order to solve Argentina's racial problems.

"An Indian is an Indian above all": Mansilla's Change of Heart

One of the aspects of Mansilla's life and work that has challenged historians and literary critics is the radical change in opinion he demonstrated with regard to the fate of indigenous peoples. While *Una excursión* argues for justice, mercy, and incorporation, by 1885 he maintained in front of the Cámara de Diputados of the National Congress that his personal experiences with indigenous peoples proved they were "organically, for evolutionary reasons, refractory to the type of civilization we are trying to make prevail and spread in all of the vast extension of these two Americas."[55] For this reason, he argued, he would vote against proposed funding for the formation of Indian colonies; anthropology had convinced him that the incorporation of indigenous peoples to civilization would not have positive results.[56] In the same session, he also called for a special law that "establishes that an Indian is an Indian above all, and that, whatever the reasons our fathers and the legislators had to declare that all men born in the territory of the Republic are Argentines, we cannot equate the Indian with the other inhabitants."[57]

In a period of fifteen years, Mansilla's thinking evolved into the near opposite of what he wrote in *Una excursión*, and he used racial determinism to exclude the Indian from property ownership, citizenship, and much-needed government assistance. These sentiments cannot be seen as merely a reflection of the time, for many of Mansilla's fellow representatives disagreed with him and Chile established a similar system of colonies one year later.[58] How could someone so willing to be seduced by barbarism in 1870 be so against it in 1885? Many factors may have contributed to Mansilla's change of heart, but I propose that his depictions of sex and gender in *Una excursión* foreshadow this change in his scientific and political beliefs. Whereas Ramón Lista's experience of sexual desire in life appears to have caused a change in thinking apparent in his texts, the Mansilla presented in *Una excursión* is a constructed narrator retrospectively telling the tale of his excursion for publication in a public forum. What we read in *Una excursión*, therefore, is a highly selected, highly manipulated account of his experiences among the Ranqueles. A critical reading of the text reveals that although he used desire to argue for equality, this desire was unconsummated and largely illusory.

Throughout *Una excursión*, Mansilla counters every expression of desire with an assertion of his own control. His relationship with Carmen, a Ranquel women, most clearly demonstrates this narrative tension. Carmen is introduced very early in the text and remains the most consistent female presence throughout. Although he barely describes her physically, he immediately presents the possibility of sexual relations between the two. Rosas has sent her as a spy, Mansilla claims, for he believed in woman's power over man. As with the other indigenous women, Mansilla depicts himself as tempted, indicating that he has had to confront all the "seductions [that] can assail the diplomacy of a frontier commander who has to deal with secretaries like my *comadre*." Nonetheless, each expression of desire is quickly countered by a negation that reinforces Mansilla's control over both the narration and the Ranqueles. For instance, here Mansilla insists that he was not "another Hernán Cortés," distancing himself from the conquistador known for having a mestizo child with his interpreter, Doña Marina (also known as La Malinche) (6).

Other scenes with Carmen further reiterate this back and forth between desire and denial. When Carmen visits him late at night, he describes the fluttering of the fabric on her chest and hints that if he were not so old, he would have assumed that she had come to seduce him (332). As in the opening passages, here again he demonstrates his (and her) interest, even while closing the door on any such possibility. Finally, right before he leaves the toldos, Carmen's sister comes to him and informs him—"with a confidential air, her eyes gleaming as those of women only gleam when a wanton thought crosses their imagination"—that Carmen is waiting for him (351). The description of her sister again raises the specter of a relationship between Carmen and Mansilla. Nonetheless, in the next breath he notes that he asked a soldier to accompany him to the meeting. He claims to have asked due to his fear of dogs, but the soldier also provides a convenient chaperone, just as the Franciscan priests did in the Villarreal toldo.

These passages highlight that even when appearing to be seduced, Mansilla believes himself to always be in control. Furthermore, he echoes his predecessors and peers in the Americas by interpreting indigenous women's interest in him self-servingly. Studies in North America and Argentina have shown that indigenous women frequently acted as intermediaries for their own cultures, "part of a diplomatic strategy to initiate trusted interaction, establish economic or political alliances, and form lasting kinship bonds."[59] White men were unable to see these motivations or actively ignored them, and instead interpreted indigenous women's interest in them as proof that they were drawn to civilized culture.[60]

In *Una excursión*, Mansilla acknowledges early in the text that Mariano Rosas had sent Carmen to seduce him in order to gain access to the details of his expedition. Despite admitting that her explicit purpose is to manipulate him, he never returns to this possibility and instead egotistically views the moments when she and the other indigenous women flirt with him as successful seduction by his civilized masculinity. Perhaps Mariano Rosas also viewed himself as victorious, thanks in part to the machinations of Carmen. Of course, Mansilla needs to deny Carmen complex agency in order to keep his ethnography together. All of his knowledge about Ranquel religion and sexual practices comes from Carmen, and she is also his "instructress" in Araucanian (219, 224). If he does not reassert his control over her, then his anthropological insights are potentially false, provided by an informant with her own motivations. As across the Americas, Mansilla's self-centered interpretations erase potential agency and cunning on the part of indigenous peoples and insist on his ultimate superiority and control.

The Mansilla-Carmen-Rosas triangle also highlights a challenge of understanding indigenous women's participation in shaping or resisting scientific-colonial projects on the Argentine frontier. My counterreading of Carmen and Mansilla's relationship as well as the other examples in this book clearly demonstrate that interactions between indigenous groups and scientists were often mediated through women's bodies and actions. Carmen uses flirtations to inform and manipulate Mansilla; Clorinda Coile was both the mother of Lista's child and his informant; and Moreno, Zeballos, Lista, and Musters were offered indigenous women as payment for services and/or as a diplomatic strategy. Although these are clear examples of previously unacknowledged indigenous agency in the development of anthropology, it is less obvious if the indigenous women were voluntary, independent actors or if their participation was dependent on the continued enforcement of patriarchal relations within the tribes. Musters notes that the indigenous widow he was to marry was disappointed when the deal fell through, but he also indicates that the tribe had encouraged her to approach him. In Lista and Zeballos's accounts, indigenous men offer them the women, and Mansilla insists that Carmen is following Mariano Rosas's orders. The limited perspective of their texts provides very little insight into the extent of indigenous women's participation, although it is apparent that in these cases, neither women's presence nor their actions necessarily implied their agency.

Finally, Mansilla's foray into interracial sex is never consummated in the text. In *Colonial Desire*, Robert Young asserts that colonialism is a desiring machine, hungry for land, resources, and people. Nineteenth-century Europeans, he claims, felt a sexual attraction toward native peoples that was contrasted with a simultaneous revulsion brought about by the specter of the interracial child that would result from such unions.[61] Colonial desire is thus a combination of attraction and rejection, just as Mansilla narrates his attraction and denial in interactions with Carmen and the other women. In *Una excursión*, Mansilla leaves behind no children, only a "godson, the future Chief Lucio Victorio Mansilla" (259). This relationship requires no racial mixing, just the imposition of a Christian name on an indigenous child in a paternalistic relationship that reflects Mansilla's perception of the Argentine government's ideal relationship with the Ranqueles. The only place the relationship between Carmen and Mansilla takes flight is in a dream in which he is the emperor of the Ranqueles and "*Lucius Victorius Imperator*, has had Indian Carmen crowned empress" (335). While the dream would seem to legitimize his desire for Carmen, the moment is surrounded by laughable images, such as Mansilla riding a glyptodon. The general absurdity of his dream means that his oneiric union with Carmen is cast as an impossibility rather than the triumphant merging of two cultures.

Thus, *Una excursión* holds both of Mansilla's positions in an unresolved tension as he works to revalue indigenous peoples while also maintaining his own position as a member of the European-Argentine elite.[62] His presentation of sex and desire most clearly illustrates this friction, allowing a glimpse into the future political and scientific positions he would forcefully assert. We should thus view Mansilla's foray into the Ranquel world not as a true conquest of barbarism over civilization, but rather through Phillip Deloria's concept of "playing Indian." As Deloria argued, temporarily dressing up and adopting indigenous garb or attitudes "offered Americans a national fantasy: identities built not around synthesis and transformation, but around unresolved dualities themselves. Temporary, costumed play refused to synthesize the contradictions between European and Indian. Rather, it held them in near-perfect suspension, allowing Americans to have their cake and eat it too."[63]

While Deloria is speaking of the actions of a North American population, as exemplified by the ethnographer Lewis Henry Morgan, *Una excursión* illustrates similar patterns in the southern hemisphere.[64] During his excursion, Mansilla enjoyed freedom from the constraints and hypocrisies of civilized life, but he did not shed his European identity or sense of superiority. He can cut his toenails at the table, declaring, "This Christian colonel might have been an Indian! What more could they [the Ranqueles] desire? I was moving toward them. I was becoming like them. It was the triumph of barbarism over civilization" and, within half a page, quote French poets and reference the Roman emperor Aulus Vitellius Germanicus (*A Visit*, 238). This duality, as well as the dialectic nature of colonial desire, allow him to admire and even fantasize about indigenous women while in the same instant negating any chance of acting on those desires. The charade was always temporary. Although intense, Mansilla's excursion only lasted eighteen days and he wrote *Una excursión* once back in the bosom of civilization. The sympathetic feelings gained from playing Indian wear off quickly, and years later Mansilla is able to argue against the Ranqueles, for, as Deloria insists, playing Indian "offered the concrete ground on which identity might be experienced, but it did not call its adherents to change their lives."[65]

Exceptional Men

Two of the nineteenth-century texts with the most positive view of Argentine indigenous peoples came from the authors who most openly or obviously admitted to sexual attraction to indigenous women. Which direction did this relationship run? Were Mansilla and Lista more positive about incorporating the Indians because they were sexually attracted or were they more likely to be sexually attracted to them because they already viewed indigenous peoples favorably? Both Lista's noticeable shift in emphasis post-affair and Mansilla's stated change of opinion regarding the Ranqueles support the theory that attraction—and thus humanizing—bears the fruits of tolerance. Regardless, the relationship is not fully determinative, for while Mansilla only plays at going Indian, Lista stays for several years, has a child, and returns to civilization only under duress.

In both cases, there is a connection between attraction for indigenous women and attraction for the freedom and exoticism of the frontier itself; this is perhaps unsurprising given the common personification of the land as feminine. *Una excursión* and Lista's texts created a voyeuristic experience that let readers live out their fantasies of going Indian and exploring another world consequence-free. Through these processes, they could temporarily cast aside the controlled rationality of civilized masculinity in order to vicariously experience the passion and disorder of primitive masculinity.[66] This desire was primarily the realm of an elite, masculine, urban community that viewed the pampas with trepidation but also as a romantic place of escape from everyday life. For these faraway readers, this world appeared exotic and interesting. It was likely a mundane reality for those actually living on the frontier.

In these desires, both Lista and Mansilla benefited from their privileged positions as white, urban males. While nomadism was one of the characteristics most frequently used to cast indigenous peoples as barbarians, Lista's impeccable pedigree allowed him to long to live like a nomad without consequence. Similarly, lower-class men who flirted with barbarism could be tainted by the encounter, but Lista and Mansilla possessed the power and unequivocally European, upper-class background that allowed them to play Indian without actually becoming one. Even Lista's affair with an indigenous woman, a situation that would be reviled if engaged in by lower-class men, posed no obstacle to his continued participation in Argentine scientific life. Of course, in both texts it is also clear that it is white men who should take the lead in interracial relationships; women partnered with indigenous men continue to only be imagined as victims.

Were Lista and Mansilla exceptions? Discussions of the erotic aspects of ethnography occur rarely. However, these types of relations appear to be somewhat common and constitute "one of the dirty little secrets of ethnography."[67] In the 1820s, Henry Rowe Schoolcraft (who Lista cites in his works), married Jane Johnston, a woman of Ojibwe descent.[68] In the last fifty years, both Kenneth Good and Hans-Joachim Heinz married and wrote about younger women native to the cultures they studied. In those texts, as with Lista and Mansilla, cross-cultural relationships become "an entry to ethnographic observations and knowledge of the 'other'" such that sexual desire is represented as a solution rather than a problem.[69] These examples suggest that Lista and Mansilla were not so exceptional, making a convincing argument for overcoming our commonly held belief that anthropologists are not sexual beings, or that they could never be attracted to populations they describe as inferior or primitive. What riches could be explored if we were to read other nineteenth-century texts with an eye to the ways that the anthropologists revealed, manipulated, or hid their sexuality as they interacted with indigenous peoples?

CHAPTER 3

Displaying Gender

Indigenous Peoples in the Museo de La Plata

AS MILITARY ACTIONS IN Patagonia drew to a close at the end of 1884, the Tehuelche caciques Inacayal and Foyel voluntarily travelled to the Villegas Fort in Junín de los Andes to reach a peaceful surrender with Argentine forces. Instead, they were taken prisoner with many members of their tribes and sent to the Retiro barracks in Buenos Aires, over one thousand miles from their home. Unsure of their future and worried their family members might be "redistributed," the caciques reached out to an old friend: Francisco P. Moreno, the explorer and scientist at the helm of the Museo de La Plata. Moreno responded immediately to their summons, then began contacting various officials to try to improve the Tehuelches' situation until they would be returned to their lands. Other interests, misinformation, and bureaucracy intervened, but sometime in September or October of 1886 Moreno's efforts paid off and the minister of war Carlos Pellegrini gave permission for the chieftains and ten to twelve members of their entourage to be released into Moreno's protection and brought to live at the Museo de La Plata, not yet open to the public.[1] There, they were studied and photographed, and worked in weaving or museum construction according to gender.[2]

The story of Inacayal and Foyel has captured the attention of the Argentine public and scholars alike due to its shocking nature and the ways in which it lays bare the close ties between anthropology and Argentine military attempts to subdue indigenous tribes. As my own writing above demonstrates, however, over time the two chieftains' names have become shorthand for the members of the whole group. For example, Mónica Quijada demonstrates how Inacayal's death in the museum and Foyel's return to Patagonia illustrated the two options available to indigenous peoples: Inacayal's death in the museum made him totally visible, exposed and literally naked even of flesh, a permanent relic of the national past. Foyel lived but was made invisible when he returned to the south, assimilated, and was subsumed into the category of the poor.[3] Their experiences are postulated as representative, but Inacayal and Foyel's high status in their tribe, close friendship with Moreno, and gender meant that the way they were treated in life and photographically was not the same as the experiences of the men and women who accompanied them.[4] These reductions—motivated by practical concerns about writing style and ingrained biases toward male leaders—reduce the diverse and gendered experiences of the Tehuelches in Buenos Aires and La Plata.[5]

Nonetheless, evidence of this complexity remains in the over forty-five photographs of the Tehuelche Indians that Moreno commissioned from the renowned English-Argentine photographer Samuel Boote's photographic studio around 1885, as well as in photographs of a subgroup of those individuals taken at a later date in an indoor location.[6] Despite a tendency to focus on the male caciques, in the photographs taken by Boote in Retiro, the gender distribution of his subjects is nearly equal. Among the adults and children over the age of seven, there are twelve female subjects and eleven males, while twenty-five of the total images depict women and twenty-one depict men. The large group shot is the only image that depicts men and women together, and each of the people in that image were also photographed individually. Men and women were photographed in very different ways, often in dialogue with the understandings of sex, gender, and race circulating at the time.

Overall, Moreno's choice of subject, the gendered visual languages developed by Moreno and Boote, and the way the images and the Tehuelches' real bodies have been preserved and remembered reiterate once again the important presence of women in the encounter between Argentine science and indigenous groups. Additionally, scholarship's focus on the Tehuelches' real victimization often depicts them as passive participants in an artistic-scientific process associated exclusively with nonindigenous culture, further invisibilizing the very people most narratives claim to be vindicating. Combined with textual records, my analysis of gender in the images demonstrates that even within the highly restrictive context of their time in the museum, the Tehuelches did exert some agency, leaving their mark on the Boote/Moreno photographs. Thus, by shifting focus from the purpose of the images to their gendered iconography, this chapter aims to demonstrate a possible escape from the dichotomy between heroic science and genocide that has troubled the development of knowledge about nineteenth-century anthropology and its cultural products.

The Boote/Moreno Images: Form and Function

Photography of indigenous peoples served a number of purposes in the late nineteenth-century River Plate. Photographers accompanied military expeditions, missionaries commissioned images to justify and publicize their works, and commercial photographers snapped shots for albums and postcards. Like their counterparts worldwide, Argentine scientists also began to experiment with the new technology. Photography and anthropology developed almost contemporaneously, and early anthropologists were eager to use cameras to record racial types around the world. In addition to presenting a relatively easy, transportable way to collect data, the presumed objectivity of the photograph seemed to legitimize anthropological practice as true science.[7]

Inacayal, Foyel, and their tribes' arrival in Buenos Aires coincided with a peak in the scientific photographic representation of indigenous peoples in Argentina. By the mid-1870s, the Argentine Scientific Society sponsored yearly photography competitions, and explorers such as Estanislao Zeballos hired photographers like Arturo Mathile and his portable camera lab for their expeditions.[8] Moreno and his museum developed similar projects. In 1878, he sent frontal and profile photographs of fifty indigenous skulls to the World's Fair in Paris; after the exposition, he donated the album to Paul Broca's Société de Anthropologie de Paris.[9] He also took his own images while traveling through Córdoba, Mendoza, and San Juan between 1882 and 1884 and participated in photographing the captured cacique Pincén in Buenos Aires.[10] Many of these images were later displayed in his museum, further underscoring their scientific value. It is not surprising, therefore, that Moreno was eager to photographically capture the men and women he encountered in the Retiro barracks and later brought to live in La Plata.

It is challenging to precisely identify the corpus of images that should be attributed to Moreno and the museum. While many texts refer to the Tehuelche photographs, there is little agreement about when and where they were taken, which images belong to the series, who the people are in each image, and why the images were taken. Even at the time, contradic-

FIGURE 1. Cacique Shaihueque (ARQ-002-011-0006) © Museo de La Plata. Photographs from the archives of the Museo de La Plata have been digitized as part of the British Library's Endangered Archives Programme, in the collection "'Faces Drawn in the Sand': A Rescue Project of Native Peoples' Photographs Stored at the Museum of La Plata, Argentina." Photographs from the Department of Archaeology can be found at eap.bl.uk/collection/EAP207-2.

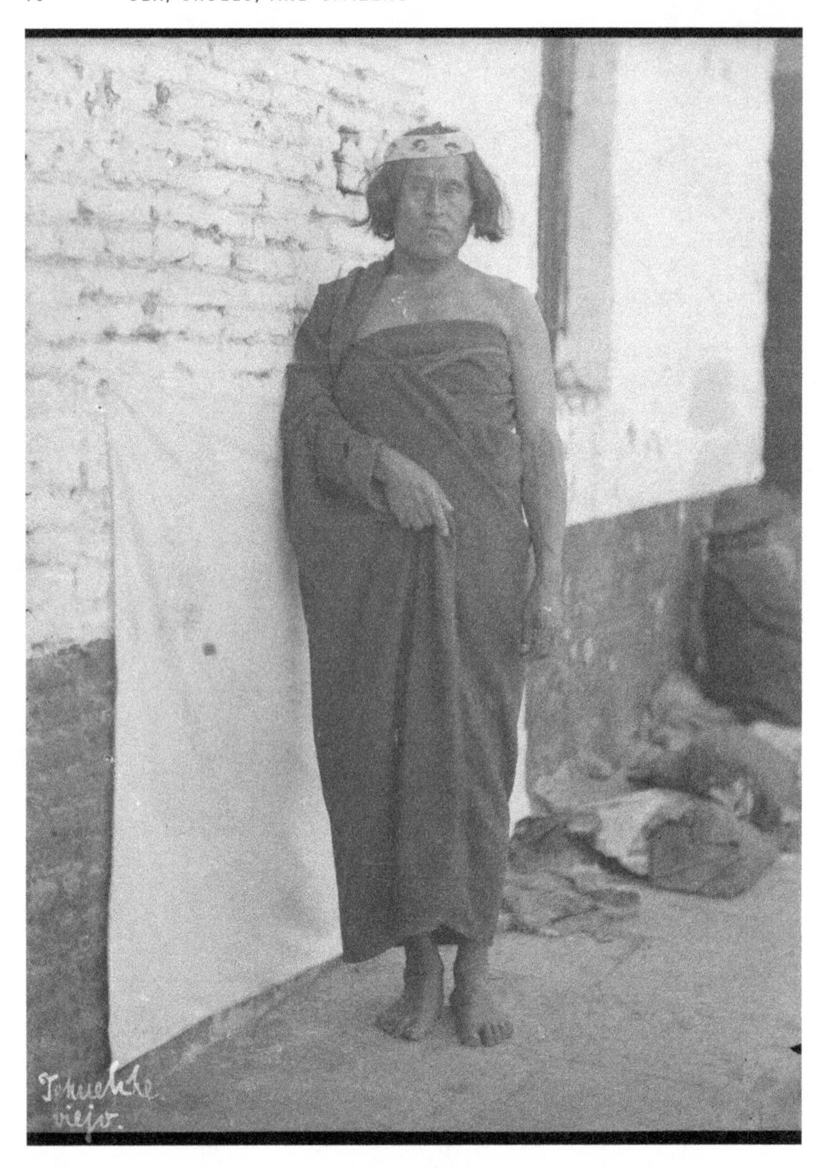

FIGURE 2. Tehuelche man (ARQ-002-014-0006) © Museo de La Plata

tions existed among Moreno's various versions of the Tehuelches' stay at the museum and third-party newspaper accounts of the same. Further complicating matters, the images were not immediately labeled nor published as a cohesive group. Photographs are material objects, and the ways that they are manipulated, displayed, and otherwise contextualized or altered over time shape the "realities" seen within the frame.[11] The scholar of these images

FIGURE 3. Inacayal, Foyel, and family, likely in the Retiro barracks (ARQ-002-006-0002) © Museo de La Plata

must thus contend with over a hundred years of intervening factors, including multiple attempts to label them, physical changes such as cropping, and the theft, misplacement, and degradation of both negatives and prints.[12]

In 2008, investigators at the Museo de La Plata received a major grant from the Endangered Archives Program of the British Library to digitize and catalog the museum's photographic collections. The resulting catalog attributes eighty-one negatives and twelve albumen prints to "Samuel Boote, by order of Francisco Pascasio Moreno, [taken] in the barracks of the Tigre Regiment (1885) and probably in the Museo de La Plata (ca. 1886–1888)."[13] While the museum groups them together, comparison of the props, backdrops, and individuals photographed leads me to believe that the collection contains two distinct series, taken at different times in different spaces of the same location. One set, with more regular stones, a higher transition on the wall from stone to plaster, and a ladder-back chair, contains images of the cacique Shaihueque, known as the "Governor" of the Southern regions (Figure 1). The other set has a slightly different backdrop and a chair with a curved back (Figure 2). These nearly fifty images include the caciques Foyel and Inacayal and numerous men, women, and children that appear both in a large group photo (Figure 3) and in individual portraits. The differing dates of the two groups' imprisonment in Buenos Aires, Moreno's own recollections, and an invoice dated July 30, 1885, from "Company of Samuel Boote" to the "Museo y Biblioteca 'La Plata'" "For taking 47 portraits of Indians with their corresponding negatives at a cost of $4 each" confirm this division of the images.[14] In this chapter, I focus on the Boote images of Inacayal, Foyel and their tribe, as well as another series of the same individuals taken in an indoor location, likely the Museo de La Plata, sometime after October 1886.

Many other Argentine photographic collections of indigenous peoples from the time focused on either men or women, and this gender specificity related to their purpose. Within images of the Conquest of the Desert, the adult subjects are primarily male: 100 percent in the Antonio Pozzo album (1879) and 84 percent in the Carlos Encina and Edgardo Moreno album (1883).[15] Encina and Moreno's album promoted the military actions of the Argentine state and so depicted indigenous peoples as savages that needed to be controlled.[16] As I have demonstrated in prior chapters, men best fit this profile, particularly through their depiction as hypersexed animals that threatened good Christian families and thus the imagined national family. Conversely, women were the subjects of most late nineteenth- and early twentieth-century postcards of indigenous peoples. As in Mansilla and Lista's written texts, in this context exoticized and sexualized female indigenous bodies were laid bare for white desires as they circulated to delight and surprise the cards' recipients.[17]

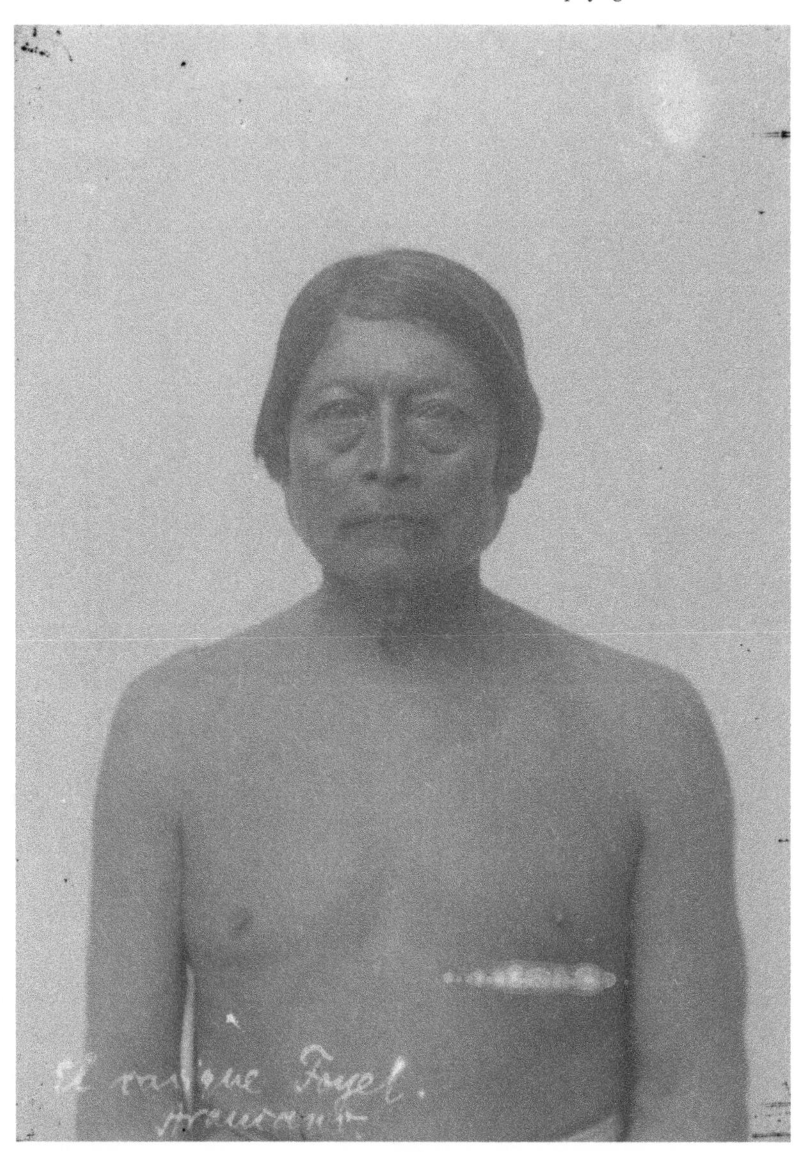

FIGURE 4. Cacique Foyel (arq-002-012-0001) © Museo de La Plata

Moreno and Boote's images, in contrast, do not adhere completely to any one aesthetic convention, frustrating scholars wishing to impose a uniform meaning on them. One of the most apparent inspirations for the Tehuelche images was anthropological photography, specifically racial-type and anthropometric photography. In these types of photography, individuals' bodies could prove their racial identification and the physical, psycho-

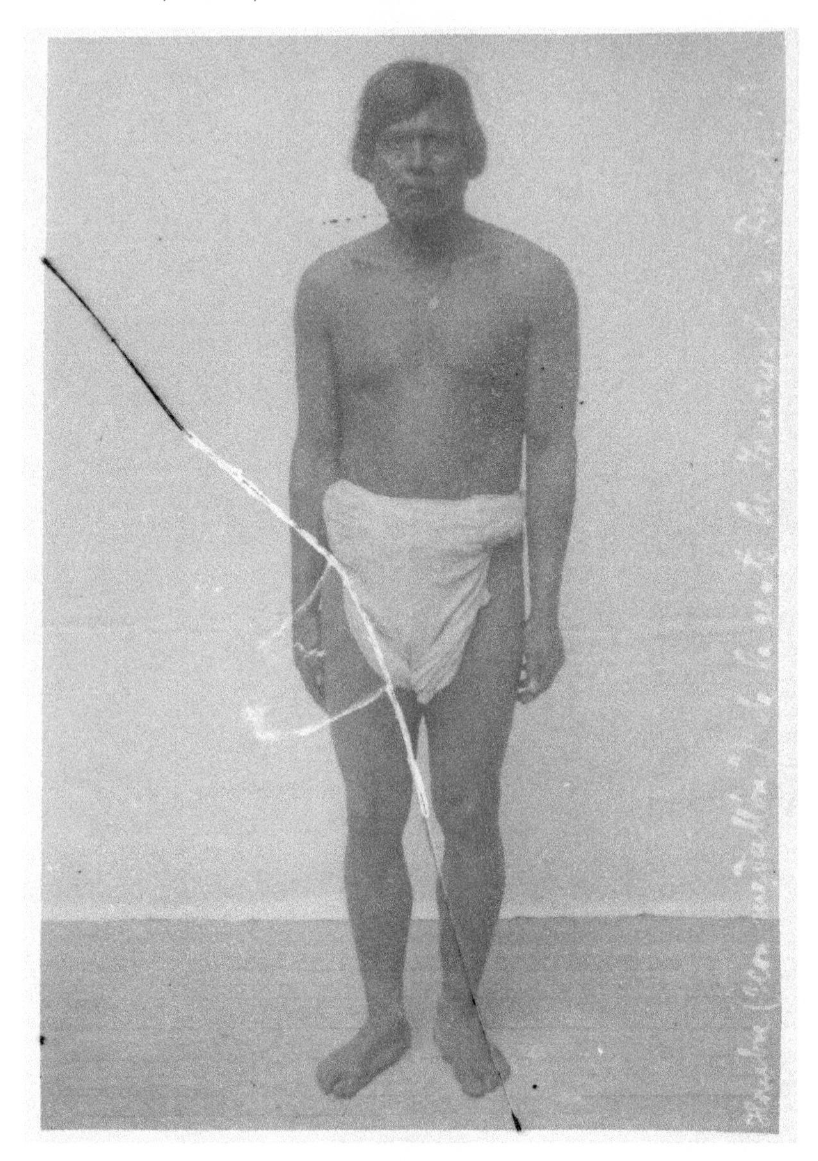

FIGURE 5. Araucanian man (ARQ-002-001-0005) © Museo de La Plata

logical, and social characteristics of their race. Marta Penhos and Máximo Farro have demonstrated that Moreno was aware of the emerging conventions of the field, many of which can be seen in the images.[18] Most of the pictures in the series are frontal and profile portraits taken in front of a neutral background (a white wall in the museum images and a white sheet in the Retiro series), and there are few props and little visual noise, allow-

FIGURE 6. Inacayal's youngest daughter (ARQ-002-021-0010) © Museo de La Plata

ing the subjects' dimensions, musculature, and other physical characteristics to stand out (Figures 4 and 5).[19]

Several of the images were later printed and hung in the Museo de La Plata's anthropology gallery, which also included skulls and mortuary masks in an effort to mimic European museum practices. This gallery also communicated with the comparative anatomy room where the skeletons of some of the images' subjects were displayed.[20] Together, the multiple representations of the Tehuelches "forge[d] a gruesome allegory of state conquest by exposing, in the museum interior, the radical exteriority of an otherness that the rationality of liberalism can only conceive as—and thus turns into—a space of death."[21] Thus, the museum became a "safe space" where racial others could be "experienced but not engaged."[22] The glass walls of the display and the images' heavy frames allowed museum visitors to see but not touch, observe but not interact.[23]

At the same time, many of the photographs fall short of a true anthropometric style, suggesting other priorities. There are no measuring devices visible in the images, nor anything used to maintain a constant scale. Despite the presence of the neutral background, the vast majority of the photos were taken at an angle to the sheet, rather than head on. The hard border between the sheet and the wall is clearly visible, and often intersects with the subject's head or shoulder. Additionally, the background is apparent in many of the images, including a stone wall, lights, buildings, and even the blurry heads of other indigenous people. These elements distract from the focus on the Tehuelche subject and render risible the attempt at anthropological photography represented by the white sheet. Finally, many of the images are labeled with not just the subjects' ethnicity, but also their proper name. These labels are at odds with racial-type photography that stressed generic racial identity over individualism, including that seen in other Argentine photo series of the period.[24]

The other major aesthetic tradition echoed in the photos is that of the bourgeois portrait and the *carte-de-visite*, characterized by full body or waist-up frontal or ¾ shots, often posed around a chair or similar object. Inacayal, Foyel, Inacayal's youngest daughter (Figure 6), and the interpreter, Rufino Vera, pose in this fashion (Figure 7).[25] Whereas racial-type and anthropometric-style photography transformed indigenous bodies into scientific objects, photographing the Tehuelches using the same techniques as middle-class Argentines could legitimize and depict them as ready and worthy of national incorporation.[26]

Indeed, in a March 1885 article in *El Diario* that summarized Moreno's meeting with Inacayal and Foyel in the Retiro barracks, he criticizes the Tehuelches' imprisonment and reiterates their contributions to his explorations, and thus the nation: "Amongst these [Indians] there is not a single one who has treacherously mistreated a white man, and if they have done it, it must have been done in the difficult struggle for life, in legitimate combat."[27] Whereas the struggle for life had previously been used to justify Argentine actions, here Moreno uses it to excuse indigenous violence. He argues that his personal experiences with Inacayal and Foyel, as well

FIGURE 7. Rufino Vera (ARQ 002-001-0006) © Museo de La Plata

as their interactions with his precursors, George Chaworth Musters and Guillermo Cox, prove that "the habit does not make the monk. Inacayal and Foyel are civilized men, hidden beneath the *quillango* or Pampean blanket." Through the article, Moreno personally vouches for the Tehuelches' worth and ability to be tamed. He finishes, "I write these lines so that *El Diario*'s readers know the end of a good tribe. . . . I repeat: Inacayal and

Foyel deserve to be protected and not confused with Pincén or Namuncurá's type. They have not killed, they have given hospitality. May they not have, then, the same unhappy end as Orkeke's tribe."[28]

The composition of many of the Moreno/Boote photographs offer proof of Moreno's argument that Foyel, Inacayal, and their tribes could be easily incorporated into the Argentine nation. Rather than attempt to depict some sort of "authentic" Tehuelche culture, the images reveal a mixture of creole and indigenous elements. In comparison to the rest of the barefoot men, Rufino (Figure 7) wears boots and gaucho clothing as he poses next to a chair. Behind him, other Tehuelches in traditional clothing sit on the ground. The juxtaposition of their blurry bodies in the background and Rufino visually testifies that the Indians were "civilized men, hidden beneath the *quillango* or Pampean blanket." With some effort, they could be transformed, just like Rufino rose from the masses. Similarly, Inacayal's youngest daughter, standing alongside the same chair, wears traditional clothing and a rosary, pointing toward her willingness to assimilate. Finally, while props are overall scarce, European-style chairs and electric lights intrude on the frame, inserting the Tehuelches into the national context (see Figure 3, far left).

While these competing aesthetics reveal Moreno's conflict over whether the Tehuelches were anthropological material or future citizens in formation, it is important that his ambivalence about this particular group not be generalized. He makes very clear in his article in *El Diario* that his defense of the Tehuelches is not a broader call for indigenous incorporation. Indeed, in order to emphasize Inacayal and Foyel's civilizability, he highlights the unsuitability of other groups: "Inacayal and Foyel deserve to be protected and not confused with Pincén or Namuncurá's type." Images of Maish Kensis, a Yámana man, and Arturo, an Alakaluf man, present in the museum in approximately the same time period, do not display the bourgeoisie elements seen in the photos from Retiro and instead hew more closely to the standards of racial-type photography. The Boote/Moreno photographs of Foyel, Inacayal, and their tribes thus are an important reminder of the ways that personal connections, contingencies, and even chance give rise to individual manifestations of broader historical trends as well as the danger of black-and-white interpretations.

Images of Indigenous Men and Women

Unlike field photography or postcards, Moreno's anthropological and propagandistic projects were dependent on visually representing both men and women. Following the naturalist tradition of identifying male and female examples of each species, late nineteenth-century scientists collected images of both men and women of the races they studied. When the English comparative anatomist Thomas Huxley wrote to colonial officials requesting anthropometric portraits of colonized peoples, his instructions specifically requested images of an adult male and an adult female of each human type.[29] Given that men and women were generally assumed to be fundamentally different, recording both sexes allowed a more complete understanding of the physical characteristics of any given race. It also permitted comparisons between races along gender lines, such as the later use the Dutch anthropologist and Museo de La Plata curator Herman ten Kate would make of the images, brains, and skeletons of the Tehuelches as he compared them to African and English men and women.[30]

As seen in the Argentine anthropologists' written texts, indigenous women were also key to the civilizing project. For the principal scientists, indigenous women's suffering while living in their current state, gendered connection to Christianity, and potential role as sexual partners for white men located them at the center of efforts to beat back barbarism. It makes sense, therefore, that Moreno would choose to capture the female members of Inacayal and Foyel's groups on film.

As in the Argentine ethnographies and political documents, in the Boote/Moreno photographs men and women's relationship to civilization is depicted as inherently gendered. Among the images of men, the primary sign of their willingness to assimilate is the creole clothing that many wear. In contrast to the furry quillangos they wore for their nomadic Patagonian existence, this lighter attire, and particularly the gaucho boots worn by Rufino, represents them as ready for indoor spaces and permanent housing. It also references the particular role Moreno prescribes for them in society. It is not the same finery as seen in portraits of the explorers and politicians of the era, but rather cheap clothing adequate for workers of the lower classes. In this sense, Moreno's imagery foreshadows the processes of invisibilization that would structure Argentine-indigenous relations in the coming centuries as indigenous peoples were integrated into regional economies as *pobres* (the poor) and *campesinos* (countryfolk), not as specifically marked indigenous bodies.[31]

The iconography Moreno and Boote used to depict indigenous women similarly reflects circulating ideas about women's relationship to civilization. Contemporary Argentine photography of indigenous peoples, particularly images taken for missionary purposes, tended to show women as westernized and ready to be incorporated into the civilized world.[32] Moreno and Boote's images follow this tradition, which also echoes the textual depictions of indigenous women seen in the anthropological literature of the time. At least four women, all from Inacayal's family, wear rosaries. In the nineteenth century, religious conversion was one of the State's primary markers of assimilation. The 1853 Argentine Constitution made indigenous people citizens, but it also tasked Congress with promoting their conversion to Catholicism (article 64, 15), contradicting article 14 assuring freedom of religion for all inhabitants. Hispanic Catholicism's pervasive association of women and religion also contributed to the sense that indigenous women would be more easily civilized than their male counterparts. Indeed, when Lucio V. Mansilla depicted a baptism among the Ranqueles, his focus was on describing the numerous women who attended and the sermon that the priest directed to "the mothers of the newly christened children, exhorting them to bring up their offspring in the law of Jesus Christ."[33] In the images that Moreno requested from Samuel Boote, the fact that only women are wearing rosaries highlights this connection, and thus the important role they could play in the civilizing process.

Like men, women are also depicted in relation to the role they would play in civilized Argentine society. A collection of pots and a tea kettle of creole origin are visible in the several photographs that depict groups of Tehuelche women. The three artifacts are in focus and in the very center foreground, allowing the viewer to clearly connect women and domestic tasks. Furthermore, the women continue to wear traditional clothing, in contrast to the men's creole garments. Indeed, Butto finds that among 689 photos of Tehuelche and Mapuche women taken at the end of the nineteenth century and beginning of the twentieth (including the images studied here), men were much more likely to be depicted in creole clothing.[34] Because of expectations for men and women, women's dress mattered less: whereas men would become civilized through work in public spaces, women's assimilation would happen in the intimate realms of the home and the heart.

Men and women's bodies were required for both anthropological photography and Moreno's interest in presenting the Tehuelches as civilizable. Nonetheless, the competing aesthetic conventions of these two discourses, particularly in relation to the representation of sexuality, creates contradictions within the photo series. Nakedness had long served as shorthand for savagery. Representing natives without clothing highlighted their undeveloped morality and emphasized the power of the fully dressed colonial observer.[35] In the late nineteenth century, racial-type photography and anthropometric photography continued this tradition by privileging the forms of the human face and body, literally stripping away elements that distracted from them. In fact, many of Huxley's anthropometric photographs featured completely naked subjects in a new manifestation of a "long-held fixation in the West on native genitals."[36]

Furthermore, US and European comparative anatomists "repeatedly located racial difference through the sexual characteristics of the female body" by observing, dissecting, and displaying African and African American women's sexual organs and practices. There is perhaps no better example of this obsession than the treatment of Sarah Baartman (the so-called Hottentot Venus), including her display in 1810 and Cuvier's famous 1817 write up of her autopsy, which displayed her labia.[37] The public discussion and exposition of black women's genitals continued throughout the 1860s and 1870s, such as in W.H. Flower and James Murie's 1867 "Account of the Dissection of a Bushwoman."[38]

There are a number of images in the Moreno and Boote series that depict Tehuelches in various stages of undress. Among the photographs taken in the Retiro barracks, there are several in which the men are shirtless. There are also four photographs of women with their capes pulled back and their tunics pulled down to bare their breasts. The later photographs that were likely taken in the Museo de La Plata are the most properly anthropometric and feature a number of full-body frontal, rear, and profile photographs of men wearing only a loincloth.

While these photographs represent an approximation to the naked photographs seen in anthropological photography at the time, there are significant differences. Compared to many contemporary series, the Moreno and Boote images maintain a certain degree of modesty. In Retiro and La Plata, both the men and women remain at least partially covered up. The male subjects' loincloths show off their buttocks but hide their penises from view (Figure 5). Indeed, the only male genitalia in any of the images is the penis of a baby in one of the group photographs (Figure 9). Similarly, the women are never photographed completely naked as their tunics continue to cover them from the waist down (Figures 8 and 9). As recorded in Herman ten Kate's 1906 journal article, the many measurements museum employees took of the subjects' bodies also almost completely ignored their sexual organs. Unlike the number of African women whose buttocks and labia were displayed in life and death, the Tehuelche men and women were granted at least that small degree of dignity.

What is more, in both men and women's photos there is evidence that they were exposed just long enough for the camera to do its work. The men's loincloth appears to be a traditional *chiripá*, a poncho-like garment worn like a skirt, that had been tucked up in order to reveal more of their legs and buttocks. The women's capes are pulled apart to show their breasts, but still fastened at the neck, allowing them to cover themselves quickly once the shutter snapped (Figure 8). Furthermore, in most of the images of the women, their breasts are partially covered

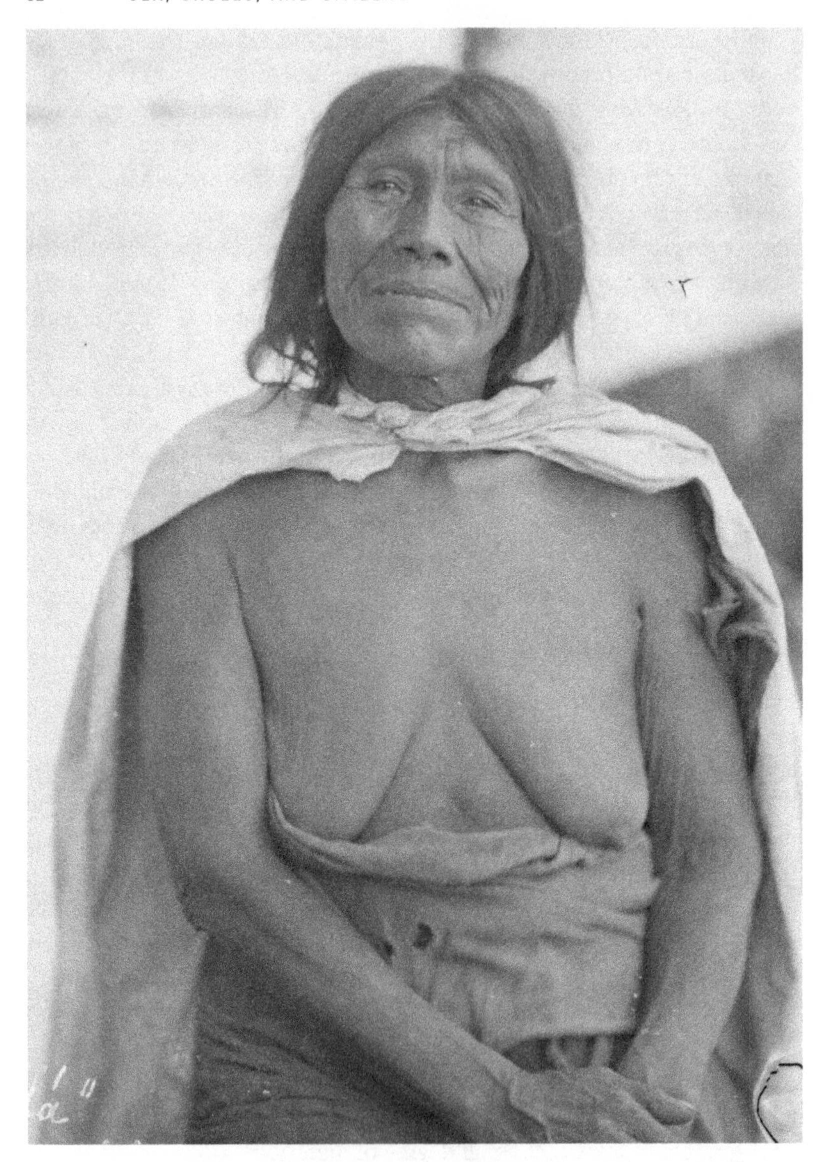

FIGURE 8. Tafá, Alakaluf Indian (ARQ-002-013-0002) © Museo de La Plata

by their hands or by a squirming child (Figure 9). For both men and women, these elements mean that the photographs stop short of meeting the scientific ideal.

Several factors likely contributed to the strange in-betweenness of the naked images, nude enough to be invasive but too dressed to properly assess the bodily composition of their subjects. First, it is possible that,

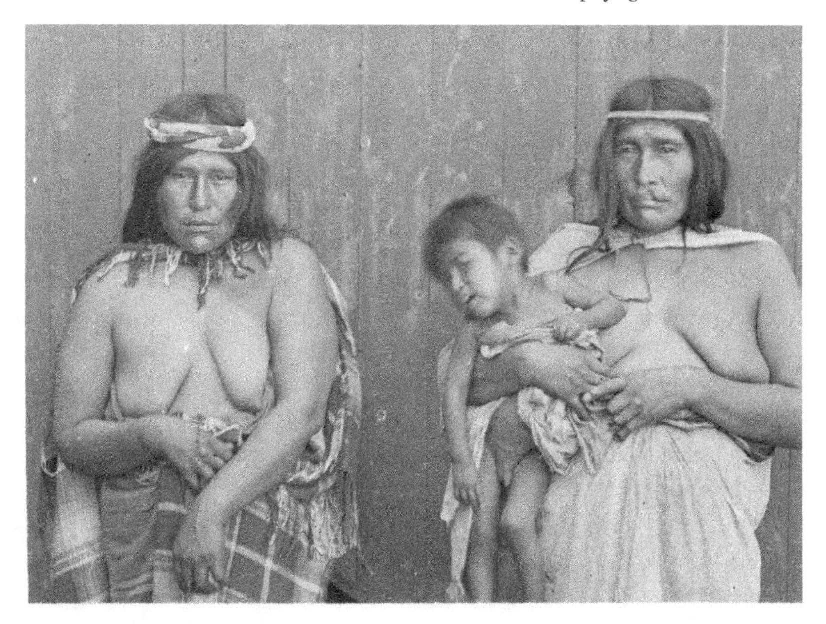

FIGURE 9. Araucanian women and baby (ARQ-002-013-0001) © Museo de La Plata

as Butto suggests, the images reflect Tehuelche habits; in Patagonia, they were not accustomed to being naked.[39] The creole scientist and photographer also had numerous reasons to deviate from anthropometric practice. In order to advocate for the Tehuelches' incorporation, Moreno needed to depict them as easily able to adopt creole gender roles. Whereas a visual trope of anthropological photography was the group of naked natives posed next to clothed Westerners, demonstrating the latter's power, in Moreno's large group image the fully dressed Tehuelches are nearly indistinguishable from the two creole guards with whom they pose.[40] Clothing could at least superficially visually collapse the differences between the colonized native and the economically dominated white lower classes.

Moreno also needed to counteract the negative depictions of Tehuelche sexuality that peppered most of his peers' (and even his own) texts. First, by covering the men's genitals and depicting them in a diaper-like loincloth against a blank wall in an indoor location, Moreno represents them as subdued, not threateningly sexual. Their shoulders are slumped, furthering this domestication. Secondly, there are no images of indigenous men and women together in family units or potential romantic pairings, echoing Moreno's earlier assertions that it was difficult to conclusively prove that "savages" knew love and that they did not "give blood ties the importance that they have."[41] It also allows the images to avoid the question of sexual reproduction by isolating men, visually reaffirming the implicit sexual goals of the distribution policy still in effect when the photographs were taken. Overall, the separation of the genders and the orderly nature of the photographs visually distances the Tehuelches and their behavior from the drunken, hellish orgies Moreno had once described in his writing.[42]

Argentine anthropologists and policy makers were also aware that assimilating indigenous peoples to creole society meant counteracting the tradition of female sexual freedom that structured many indigenous cultures.[43] The fully dressed and nude photographs of Tehuelche women explicitly reject the sexualization seen in Indian postcards and the anthropologists' depictions of native society as well as the sexual desire that structured Mansilla's text and Lista's life. When women are photographed dressed, they wear traditional Tehuelche clothing consisting of a tunic that wrapped around the woman's body and covered her from the underarms to the ground, as well as a blanket-like cape attached at the throat and concealing the arms and shoulders. These two garments, designed to provide warmth in the Patagonian cold, drape the women from neck to toes, completely obscuring the forms of their bodies. Additionally, in the images their garments are dirty or worn out, making the women appear more like prisoners than sex objects (Figure 8, Figure 10).

In most of the nude images, sexuality is similarly dismissed. Three of the four women depicted are among the most elderly, a group much maligned by the Argentine explorers. In his writing, Lista complained of how the insufferable melancholy songs of the old women disturbed his sleep and lamented that the beautiful young Tehuelche women inevitably "acquire in their old age a repugnant ugliness."[44] Moreno was even more passionate in his dislike of elderly Tehuelche women, calling them "diabolic witches, revolting monsters."[45] He later wrote that "There is nothing dirtier or more repellant than those old women" and described their small eyes, greasy, hard hair, fallen breasts, and unbearable stench with scorn.[46] The age of the women in the photographs depicts them as outside of sex and thus not a threat to creole norms regarding sexual relationships and decency.

The fact that Moreno represents the Tehuelches as already undergoing the process of assimilation presents another roadblock to the fully nude style of photography his anthropological projects might otherwise demand. As demonstrated by the rosaries worn by the women of Inacayal's family, at least some of the Tehuelches were already Christians, or at least photographed to suggest that they were. Throughout the history of Spanish-indigenous relations in the Americas, acceptance or rejection of Christianity was used as one of the primary determinants of the treatment the natives received. Sixteenth-century conquerors read the *Requerimiento* (Requirement) to indigenous peoples before battle, warning them that if they accepted the Christian god and consequently the rule of the Catholic Kings, they would be spared; in contrast, rejection of those conditions would justify theft, rape, and even death. The potential to be Christianized was similarly used to protect American indigenous peoples from slavery and justify the importation of African slaves. In the nineteenth century, Argentine anthropologists like Mansilla would also argue for the incorporation of indigenous peoples as fellow children of God.[47] In similar fashion, the women's rosaries might have precluded them being photographed completely naked. Or read more cynically, maintaining some sense of modesty might have allowed the images (and Moreno himself) to escape censure from a Catholic church already uneasy with the materialist ideas of scientists as well as their treatment of dead bodies.

Several other extra-scientific factors likely further mediated the uneasy relationship to indigenous nakedness in the images. Most of the European and US anthropologists photographed colonized strangers. In contrast, Moreno had a personal history with the tribes. He had lived among them and claimed to owe his life to Inacayal's willingness to intercede with Shaihueque on his behalf.[48] In his newspaper article describing the encounter in Retiro, he highlighted those ties: "today, I have seen those who provided my men with guanaco and ostrich meat daily, those who warned me of the Mapuches' stalking and advised me to be suspicious of them, and among the poor women I have recognized those who searched for strawberries for the white chief."[49] T. H. Holdich's 1920 obituary for Moreno published in the *Geographical Journal* highlighted the Tehuelches' dual role as anthropological subject and friend, describing how Moreno walked through his museum. He would greet the indigenous skeletons by name "and with an airy wave of his hand inquire after his [the skeleton's] health and describe in what way he was especially useful to science."[50] These affective ties likely further tempered his approach to the photographs, minimizing somewhat—but not entirely—the intrusive desire to photograph the indigenous subjects in a state of undress.

Finally, critics have generally read the Boote and Moreno photographs as moments of anthropological violence.[51] Indeed, they were taken in a context of highly unequal relationships. As Huxley discovered when he attempted to obtain images of people in the British colonies for anthropometric study, prisoners made excellent subjects due to the fact that they were a group with little power to refuse.[52] Likewise, the Tehuelches were in Buenos Aires involuntarily due to their surrender at the end of the Argentine-Indian wars. Although not charged with any crime, they could not return to their homelands and were living in less than ideal conditions. Moreno represented not just a friendly face, but also a possible lifeline as the Tehuelches anticipated that his connections might enable them to be freed. Moreno did nothing to disabuse them of this notion, and it is possible that they submitted to the photographs because they had no other choice or because they thought that pleasing Moreno would improve their chances of returning home.

Nonetheless, recent research has demonstrated that even within such unequal power structures, photography is a negotiation between the photographer and the photographed. Although prisoners may not be able to refuse being photographed, they may have the ability to choose their dress, objects, position, as well as their attitudes toward photography.[53] This tension is clear in one of the Boote photos depicting a group of women, one of whom holds a baby. Despite the photographer's best efforts, the baby cannot be controlled and squirms. The resulting blurry image is a powerful reminder of the multiple wills that come together to create photography.[54] Thus, without discounting the importance of asymmetric relations of power and the very real violence that anthropological photography imposed on indigenous peoples, we can also allow that photographs record the competing interests of both the photographer and his or her subject(s).[55] This approach acknowledges that there may be contradictions even within a single image, breaking with dualistic narratives.

Nakedness has historically been a point of tension between anthropologists and their subjects worldwide. Huxley's photographic suppliers complained of the near impossibility of getting colonized peoples to agree to be photographed naked.[56] In the United States, Franz Boas's photographs of indigenous people from the Northwest are similar to the Boote/Moreno images in how they present the tension between anthropometric composition and the subjects' full dress. Matthews attributes this disjunction to either the indigenous subjects' refusal or Boas's desire to maintain the friendly relationships he had developed with them, both probable factors in the Argentine images as well.[57] It seems likely, therefore, that the Tehuelches' pulled back capes and tucked up *chiripás* may have been compromises, allowing the indigenous subjects to limit the extent and duration of the invasion of their privacy.

The identity of the women photographed topless indicates another moment of negotiation behind the images. The wives and daughters of the two chieftains were always photographed fully clothed; at most, their shoulders are revealed. In contrast, the four women who are the subjects of the nude images came from the lower ranks of the tribe. One is Tafá, an Alakaluf Indian who was Inacayal's servant (Figure 8). The other three women are simply labeled "Mujer Tehuelche" (Tehuelche woman) or "Muchacha araucana" (Araucanian girl), with no record of their names or relationships to others. Their relative unimportance within the tribe itself is further indicated by the fact that they were not part of the smaller group later taken by Moreno to the Museo de La Plata. This class divide with regard to what are ostensibly the most intrusive photographs suggests that while the Tehuelches did not have the power to completely resist being photographed nude, they may have been able to negotiate who was photographed based on group hierarchies.[58] They also remind us that which bodies are photographed and used as examples of racial type has less to do with whether or not they are a "good" representative of type and more to do with who is willing (or can be forced) to be photographed. Thus, stereotype and anthropometric measures are driven by a degree of chance, a randomness that may have long-term consequences for our understandings of race and racial science.

As throughout this book, the photographs Moreno paid Samuel Boote to take of the Tehuelches point to the centrality of women for nineteenth-century scientific projects. While we can never know Moreno's true intentions with the images, their composition and his textual accounts of the time suggest an array of purposes including anthropology, propaganda, and even documentation of personal friends. For each of these purposes, women's bodies were as important as men's, and it is difficult to understand one without the other. In relation to ongoing dialogues about indigenous sexual practices and interracial possibilities, Moreno's photographs are most interesting for what they do not show. In contrast to the near constant specter of sex on the edges of the anthropologists' written texts, Moreno's photographs seem to make an intentional effort to avoid the topic, depicting indigenous men and women as strictly segregated by gender and asexual. Once again, scientific racism and its applications were mediated by a number of extra-scientific factors, including religion, class, personal relationships, and gender norms.

Recuperating Indigenous Women's Experiences

Anthropologists' writing demonstrates that men and women interacted differently with both Argentine scientists and the colonial enterprise in general. Sexual desire, homosocial bonds, scientific goals, and creole and indigenous gender norms pushed indigenous people toward different attitudes and conditioned their reception by the white Argentines. These shaded behaviors were also recorded in anthropologists' descriptions of their experiences photographing indigenous peoples across Latin America. In Argentina, the head of the anthropology department at the Museo de La

Plata, Robert Lehmann-Nitsche, found that the Selk'nam women in Buenos Aires for the National Exposition in 1898 were reluctant to be photographed. Lehmann-Nitsche turned to their husbands and used them to change the women's minds.[59] In Brazil a few years later, Canela men were also more willing to be photographed by German anthropologist Wilhelm Kissenberth than the women of the tribe.[60] These two instances, representative of many more, demonstrate that men and women had different beliefs, interests, and relationships to the anthropological photographers, who were always men, and that photographers both exploited and relied on gender norms.

Intriguingly, the only written comment Moreno made with regard to the Tehuelches' relationship to photography was about the women, not the men. In the article published in *El Diario*, Moreno writes that he showed the Tehuelche women a photo album titled "mujeres amigas" (women friends) that included a picture of his sister. The women were fascinated by the album and were most surprised not by the images themselves, but by the fact that Moreno had female friends, given that (according to Moreno) women in the toldos were always slaves.[61] This episode occurred just days or weeks before the Boote photographs were taken. Might Moreno have shown them the album in order to overcome their doubts about being photographed? Could they have even requested some of the images, perhaps explaining the purpose of the images which do not reflect the conventions of anthropological photography?[62]

Indigenous men and women's different interactions with photography and photographers are also manifested in the images themselves. We have already seen how Moreno and Boote used clothing and props to highlight Tehuelche men and women's unique relationships to civilized society. Other clear gender differences structure the photographs. Men are always photographed individually and more of their images reflect the racial type and anthropometric styles. In contrast, the significant number of individual photos of Tehuelche women are accompanied by at least five images in which several women appear together, breaking with the photographic conventions of racial science. Furthermore, many of the pictures depict women and children in the same frame, while men are never shown in close proximity to the youngest Tehuelches.

Many of these artistic choices echo the representations of indigenous gender roles in contemporary anthropological texts. As previously mentioned, indigenous men are always depicted alone, even when they are with others in the frame. In the large group shot, the only one in which men and women are depicted together, the men stand straight up and down with their hands by their sides or loosely clasped in front of them (Figure 3). They stand individually or in small groups, but with only incidental contact between those standing together. The viewer's eye goes straight to the men's unsmiling faces, as they are elevated above the women, spaced apart, and thus easier to distinguish. They are also in front of a light, relatively uniform background. This positioning and posing echoes the stereotypes found in the written texts of indigenous men's antisocial nature and also Argentine conceptions of man as a public individual.

In contrast, the visual representation of the women echoes the textual choice to often refer to the women and children as *chusma*, the nameless, uncountable masses. In the large group photo, the women sit on the floor, touching in an unbroken, multileveled line from one side of the frame to the other. They are dressed far more uniformly then the men, giving the impression that they all blend together. This confusion of bodies is further blurred by various blankets, quillangos, and children heaped on and around them.

Furthermore, in the photos of the women, there is physical contact between nearly all the subjects as they intentionally lean into each other. In one, a young girl wraps her arms around a sleeping child, while a slightly older boy sitting on the floor leans into her legs (Figure 10). The two older women in the frame stand so that their shoulders appear to be touching. Inacayal's youngest daughter leans into the woman on the right, who protectively places her hand on the girl's forearm. The contact between the two extends the full length of the young girl's body as she leans her head into the folds of the women's tunic. The woman on the left holds a young child, while an older girl hides somewhat behind the two, again with full body contact. To the far left, Sayeñamku's daughter sits on a stool, with the image to her left illegible due to damage. In contrast to the isolated individualism of the men, this composition visually represents women as part of the masses and participating in a generally more supportive community. In this, the images echo the frequent written descriptions of two or more indigenous women sitting close together, creating elaborate hairstyles, applying makeup, or even picking lice and other bugs from each other's hair.[63]

Strikingly, there are no images of indigenous men with children, whereas there are many in which women are touching, holding, or otherwise interacting with children ranging from infants to approximately eight years of age. These images continue a long tradition of associating indigenous women with motherhood. In the seventeenth and eighteenth centuries, illustrated maps depicted indigenous mothers with their children.[64] Nineteenth-century texts, as shown in Chapter One, similarly connected women and their offspring, highlighting the mother's key role in physical and cultural reproduction, including the persistence of language and tradition.

FIGURE 10. Group of women and children (ARQ 002-006-0005) © Museo de La Plata

In Moreno's written texts, he stressed maternal involvement while minimizing Tehuelche men's role. "Motherly instinct is the same in all social spheres and all races, from the humble Fuegian woman to the civilized European," he insisted, but Tehuelche men abandoned their wives from the time they became pregnant until the child was at least one year old.[65] This divide is reaffirmed by the images, for the men and children do not exist within the same frame as if to suggest completely separate spheres of existence. In the one shot of the entire group, the seated women create a physical border between the men and the children, reaffirming their control over the latter by holding them while also defining a male space outside of these domestic interactions.

It might be argued that the separation of men and children was the Tehuelches' choice, not Moreno's, thus reflecting their ideas of gender instead of those of creole society. Nonetheless, while other contemporary anthropologists agreed with Moreno's association of women and children, many of them depicted indigenous men as doting spouses and fathers. In *At Home with the Patagonians*, published approximately fifteen years before Moreno's images, George Chaworth Musters wrote that the Tehuelches' finest trait was "their love for their wives and children."[66] Several years later, Ramón Lista argued that a Tehuelche father ruled the family with an "affectionate authority, and hardly ever . . . punishes his wife and children;

to the contrary, he is weaker with the latter, spoiling them to the same point as the most affectionate of civilized fathers."[67] This father-child bond was also noted by the Salesian missionaries who claimed that fathers would die before being separated from their children.[68] These statements contradict the strict association of children with mothers present in the images as well as in Moreno's earlier writings, making it less likely that the Tehuelches had chosen those formations and poses.

While Moreno was probably responsible for the decision to separate the men and children, other aspects of the Tehuelches' gendered interactions with Moreno's scientific endeavors point to additional spaces where they could exert some agency even under highly restrictive conditions. As already seen, the identity of the nude subjects and the subtleties of pose likely reflected small moments of power. Where else might they have slightly altered the photographic product?

Viewing the images through a gendered lens, one notices the great number of small children, rarely mentioned or identified in texts from the time or the present. Some are just babies or toddlers, while others appear to be approaching puberty. Moreno and his assistants' apparent lack of interest in the intimate details of the Tehuelches' encounter with science has left us without the materials needed to understand their experiences, but the very presence of the children raises a number of questions whose unknowable answers point to potential moments of agency. The young children in the images make one wonder if any women gave birth while they were in the barracks or the museum. Such births were not unheard of: we know, for example, that in November 1898, a Selk'nam woman gave birth to a little girl while in Buenos Aires for the National Exposition.[69] The youngest child in the Boote series appears young enough that he could have been born during the Tehuelches' journey from Tecka to Buenos Aires. Were there others? If there were, what were the birthing, breastfeeding, and child-rearing experiences like for the Tehuelche women far from home? If there were not, how did they control reproduction? Was sexuality limited by the fact that the entire group appears to have shared one living space in the barracks and later in the museum and/or by the stress of the encounter? Or did they actively choose not to have children under such circumstances?

Similarly, Lista describes the Tehuelches' ceremony celebrating a girl's first menstruation (*enake*) and its symbolic importance for the tribe in several of his texts.[70] In the first few pages of his travel account, Moreno describes a drunken orgy celebrating the same milestone.[71] Several of the girls in the images appear to be around the age of puberty. Would the Tehuelches have been permitted to continue this ceremony in Bue-

FIGURE 11. Foyel's wife with Inacayal's youngest daughter (AFO-002-001-032-0005)

nos Aires and/or La Plata? If they were, how did it shift due to the new environment?

A related question is whether the Tehuelches' greatest fear—that their families would be separated—came to pass.[72] We do not know the names of all the more than twenty individuals who were taken to Retiro, nor do we know the exact composition of the smaller group of twelve to fifteen that was taken to La Plata. Our information is especially scanty for the children, who were largely unidentified and were generally not photographed individually. Were any separated and redistributed to creole families in the city? What happened to Inacayal's children after he and his wife both died in the museum?

Additionally, the images clearly show women from different family groups caring for the same children, such as in the photo of Foyel's wife with Inacayal's youngest daughter (Figure 11). In the image, Foyel's wife and the girl lean into each other, appearing as if mother and child. Moreno himself noted in *Viaje a la Patagonia austral* that the Tehuelches frequently used the term "brother" or "nephew" to refer to those to whom they had no blood ties, suggesting looser and more extensive familial relationships than in creole culture.[73] What sort of familial or friendly ties were preserved or broken as the Tehuelches moved from Tecka to Retiro to El Tigre to La Plata and then back to Patagonia? We cannot definitively answer any of these questions, but simply by raising them we can imaginatively enrich the historical experience of the Tehuelches, restoring dimensions to a narrative that has been repeatedly flattened by dichotomous ideas of heroic progress versus genocide. The lack of information about these topics is not proof of their unimportance but rather of a strong historical bias against the details of women's lives and the intimate aspects of history.

Moreno's letters and the memories of Emile Beaufils, a doorman at the museum who worked as an assistant during the period the Tehuelches were present, provide more concrete evidence of Tehuelche attempts to mold their circumstances to fit their needs. As in the other cases, the nature of these efforts differed for men and women. In their correspondence, Moreno and Florentino Ameghino, the museum's sub-director, lamented the Tehuelche men's indolence. On August 2, 1887, Moreno wrote Ameghino: "Make Inacayal and Foyel bring in the iron bars that are to the side of the main staircase, they can also help the workmen bring in the materials. It is not advisable to leave them idle."[74] Ameghino responded just a few days later, complaining that very little material had been brought in, for "with respect to the Indians you must remember that we have not been able to obtain anything from them for a while. Not even by eliminating their cigarettes and diminishing their rations have we been able to make them bring in a single bar."[75] As when they lived in Patagonia, indigenous men continued to be defined by their perceived laziness.

In contrast to the men's inaction, Beaufils's memories, compiled twenty years later by Herman ten Kate, depict women rebelling against Moreno's projects through concrete actions. He presents the women as much more willing to interact and answer the scientists' questions, although he complains that they change their answers to the same question from day to day. He also recounts with frustration that the women took the weaving materials provided by the museum and then snuck into the city to sell their textiles, using the money to buy alcohol for the men.[76] As in the case of the men's indolence, Beaufils presents these moments as evidence of the Tehuelches' savagery and the suffering the scientists endured in the name of civilization and scientific progress.

Read in another fashion, however, these episodes are examples of Tehuelche efforts to shape their own experience in accordance with their needs. Men's refusal to act frustrated scientific projects as the men declined to help build the very museum that would one day be the only grave for Inacayal, his wife, and several others. The women's multiple answers are not a sign of unintelligence, but rather an intentional challenging of their interrogators. When they went into town to sell their textiles, the women were preserving the same gender roles and customs noted by explorers of the Pampas and northern Patagonia.[77] They also used the weavings to buy for the men the very products being withheld in efforts to make them work, furthering reducing the possibility that they would give in and help build the museum. While certainly frustrating for Beaufils, these moments also demonstrate women's ability to take advantage of the new environment, using materials meant to exploit them to achieve their own goals. Viewed as a sign of women's agency and resistance, the episode subverts the narrative of indigenous peoples as passive victims or ungrateful rescues in order to highlight indigenous adaptability and persistence.

As in the case of Mansilla's interactions with his Ranquel informant Carmen, these episodes additionally demonstrate the entangled loyalties of race, gender, and class. Like Carmen, the Tehuelche women were working against the Argentine scientists' projects in order to protect their own ethnic groups' interests. Again, however, it is difficult to determine to what degree their actions constituted indigenous women working independently. Did their refusal to cooperate with white men also imply a freedom from indigenous men, or were their acts of racial resistance carried out under patriarchal structures? Some moments, such as their ever-changing answers, might constitute moments of true agency for indigenous women. The sale of textiles is more ambiguous, especially given that Beaufils notes that they used the money to buy alcohol for the men, not products for themselves or the children. And in the case of Tehuelche negotiation of the topless photographs, it is likely that indigenous men negotiated with white men about how indigenous women's bodies could be exposed and used. This episode also points to the fact that not all indigenous women's experiences were the same. Even as most were treated horrifically, women connected to powerful men enjoyed protections that others did not. Indigenous women's participation in scientific encounters thus existed along a continuum ranging from agency to continued treatment as passive objects whose bodies were a currency to be exchanged.

Death and the Museum

Photographic images were not the only way that Tehuelche men and women were displayed in the museum, as the bodies of those who died between 1886 and 1888 were prepared and displayed in the anthropological galleries. About a third of the Tehuelches died in that period, with a frequency that alarmed even contemporary observers.[78] On September 27, 1887, an article in the La Plata newspaper *La Capital* denounced the deaths of "a woman, daughter of a cacique" (Margarita) on September 23, a seven-year-old girl the next day (likely Inacayal's youngest), and Inacayal himself on the twenty-sixth of the same month. The wording of the article implies that the museum staff played a role in hastening their deaths, and the anonymous writer further accuses the museum of failing to secure postmortem inspection and death certificates, as well as illegally burying the bodies on museum grounds. Most graphically, the paper alleged that "They are CARVING UP the cadaver of this human being [Inacayal], in the very museum, at the time we are writing (11 a.m.)" and expressed fear for the safety of the other Tehuelches.[79]

Moreno did not deny the allegations, but instead justified his actions as necessary for the progression of science and thus the nation.[80] He claimed that the president of the Hygiene Council gave him verbal permission to dissect the bodies and bury "the remains unnecessary for the anatomical study of the bodies" on the premises. Such actions, he insisted, were absolutely necessary,

> Given the exceptional interest these dissections would have for anthropological science, for being the last representatives of races that are going extinct and which have not yet been studied. . . . Not having dissected those cadavers would be a sign of real backwardness in the scientific movement of the day, since we would have lost extremely valuable materials of study that will contribute so much to the exact knowledge of the ethnic constitution of the American races, and above all when it is precisely the Museo de La Plata that is destined to be the center of this type of research.[81]

FIGURE 12. Sayeñamku's daughter (AFO 002-001-032-0002) © Museo de La Plata

Moreno further insisted that it was in the museum's best interests to ensure the longevity of the Tehuelches. Their deaths were not the result of malfeasance, but the natural result of illnesses acquired in the contact between civilization and barbarism, exacerbated by harsh treatment at the hands of the Argentine military. Moreno, like Lista in other contexts, repeats the vanishing Indian hypothesis to order to wash his own

FIGURE 13. La china Carmen (The Indian woman Carmen). From J. Bouchet's 1890 edition of Lucio V. Mansilla's *Una excursión a los indios ranqueles*. Digitized by the Agencia Española de Cooperación Internacional para el Desarrollo. Biblioteca. bibliotecadigital.aecid.es/bibliodig/es/consulta/registro.cmd?id=428.

hands of blame. The fact that four out of five of the dead in the museum were women also confirmed a scientific belief that, as Lista claimed, "the Tehuelche woman resists less in the struggle for life."[82] Although better suited than men for civilization, she was also considered more fragile and thus more likely to succumb to the diseases encountered in the clash of cultures.

After their deaths, the Tehuelches were, as *La Capital* alleges, dissected and displayed in the museum's anthropology gallery until the 1940s.[83] As in the images, in these three-dimensional displays of indigenous bodies, gender played an important role in the construction of ethnicity and racial comparisons. Most of the photographs were labeled only with the individual's name (if known), ethnicity, and sex. As skeletons, they were similarly labeled by sex and ethnicity, and these markers were also compiled in Lehmann-Nitsche's 1911 guide to the museum's anthropological collections. Even ten Kate's study of indigenous brains specifically separated out women (Tafá and Margarita) and men (Inacayal) and compared them to women and men of other races.[84] Of all the rich details of a person's experience, in the museum's photographic and physical memories the complexity of life was reduced to race and sex, as if these were the two fundamental and mutually reaffirming vertebra of identity.

Elizabeth Edwards has asserted that in anthropological photography, "individuality is subsumed to type, for by definition individuality is incompatible with 'type.' 'Types' were very seldom named or identified beyond the very general; tribe, place of origin, or trade for example."[85] In the case of the Museo de La Plata, trade and place of origin are rarely inscribed on the images or skeletons themselves. Sex, however, is inextricably intertwined with ethnicity and nearly always inscribed on indigenous bodies, occasionally even marked twice by the gendered ending of Spanish nouns of ethnicity, and the redundant inclusion of "woman," "man," or "girl" on the label.

The image labeled "Araucanian Girl. Daughter of Sayeñamku" is emblematic of the persistence of sex above all (Figure 12). Although Sayeñamku's daughter was Araucanian, her image was used to illustrate Bouchet's 1890 edition of Mansilla's *Una excursión a los indios ranqueles* as well as Gillies's 1997 translation of the same text (Figure 13).[86] In this image recycling, she was stripped of the relationship to her father and even her ethnicity, as Mansilla's text is concerned with the Ranquel Indians of the pampas, not the Indians of northern Patagonia. In Bouchet's text, she is labeled "The Indian Woman Carmen," and her clothing was altered to reflect the style of the Pampa Indians. The same retouched image is used in Gillies's text, where Sayeñamku's daughter is labeled "An unknown Ranquel woman." What persists over time is her indigeneity, in its most general form, and the fact that she was a woman. The fact that Bouchet chose an image of one of the few young, conventionally attractive Tehuelche women in the set to represent Carmen also underscores the importance of flirtation and desire to Mansilla's scientist-informant relationship with her, as analyzed in Chapter Two.

The third piece of information sometimes included on the museum's labels was name. This aspect once again reveals a notable gender difference. Six of the men in the photographs were named, while Margarita, Saká, and Tafá are the only three women labeled by name. Even the chieftains' wives were only identified as "wife of . . .", and except for Saká, Inacayal's children are only labeled as "daughter." As in scientific texts, women were less frequently the subject of sustained anthropological interest; even when they were, it was often in function of their relationship to powerful men.

The different rate of naming of the genders has had important impacts in the intervening years. Since the late 1990s, several groups have worked to return the bones of Argentine indigenous peoples in the museum's collections to their communities of origin. In this process, named individuals are far more likely to be restituted due to the ease of identifying them and finding descendants, as well as the ability to craft emotion-producing narratives about men and women with known backstories.[87] The fact that the women were less frequently identified by proper name means that it is less likely that they will be returned to their communities. The gender divide of successful restitutions in Argentina until the present day supports this conclusion.[88] From the field, to photos, to museum displays, men and women were treated, represented, and remembered in very different ways.

In addition to demonstrating the odd legal and social position of indigenous peoples, the imbrication of race and gender, and moments of potential indigenous agency, the La Plata museum displays of the Tehuelches and their images were part of efforts to diffuse scientific thinking to a wider audience. Moreno hoped his museum "would contribute to the general education of the inhabitants of the province of Buenos Aires," and he laid out plans for public lectures and portable *gabinetes* (cabinets) that would take natural history directly to provincial schools.[89] Most of these projects were never carried out, but the museum did develop a large number of frontward-facing displays.

Although the Tehuelches all died or returned to Patagonia by the time the museum officially opened to the public, their preserved skeletons, mortuary masks, and photographic images greeted visitors to the museum until at least the 1940s. Moreno claimed approximately fifty thousand visitors per year during the 1890s, each of whom had intimate access to the remnants of the Tehuelche citizens who became museum pieces.[90] Through the carefully designed museum displays, visitors could encounter pieces of Argentine history and understand how these indigenous peoples fit into a larger story about the progress of humankind and, more specifically, the nation. This turn toward the public can be seen in the next chapter's discussion of fiction that aimed to untangle scientific mysteries relating to Argentine indigenous peoples while also appealing to a wider lay audience.

CHAPTER 4

Degenerates or New Beginnings?

Theorizing Racial Mixture in Fiction

On July 28, 1885, the naturalist and fervent Darwinist Eduardo L. Holmberg gave a talk celebrating the thirteenth anniversary of the Argentine Scientific Society's foundation. Titled "La noche clásica de Walpurgis" (The classical Walpurgis Night), his study of Goethe's classic German play *Faust* is a seemingly odd choice to honor the beginnings of a South American scientific society. Holmberg justified his subject matter by arguing that Goethe made the first steps toward discovering comparative anatomy.[1] Furthermore, he claimed, Faust was a "scientific-literary" masterpiece in which Goethe anticipated the evolutionary ideas later developed by Darwin and Haeckel.[2] For Holmberg, it was precisely Goethe's knowledge of both science and literature that permitted him to uncover and transmit the secrets of nature. In the content and context of his speech, Holmberg elaborated a convincing argument for the interdependency of literature and scientific discovery.

Holmberg was not alone in his certainty that the disciplines could cross-fertilize. In order for the racial projects the Argentines developed in scientific texts to gain purchase, they needed to tap into the collective imagination of the nation. To do so, Lista, Zeballos, Holmberg himself, and others would turn to fiction.[3] Fictional narrative was an ideal means to represent hegemonic identity projects, and the resulting stories "seduce[d] readers into national belonging."[4] To accomplish this, the authors related stories of romantic relationships, most of which crossed racial lines. Indeed, as Doris Sommer first noted, nineteenth-century Latin American readers were accustomed to reading the romance of nineteenth-century novels as allegories of national identities.[5]

In this chapter, I study the fictional novels of Zeballos, published between 1884 and 1888, as well as two epic poems: Holmberg's *Lin-Calél*, begun in 1885 and published in 1910, and Juan Zorrilla de San Martín's *Tabaré* (1888). We have already seen the nonfictional work of Zeballos in Chapter One, and Holmberg's influence in spreading evolutionary thought in Argentina is undeniable. Zorrilla de San Martín is somewhat of an outlier, as he was Uruguayan, a poet, and a politician. Despite this uniqueness, it is important to include his work because of its very close relationship to Holmberg's. The two poems have many similarities of plot: both are set in the colonial period of their respective nations, highlight the conflict between the Spanish and indigenous groups, have a mestizo protagonist, depict prohibited romantic relationships, and end with death. Furthermore, both poems were written specifically in the context of the debates over Darwinism in the River Plate, and Holmberg's poem *Lin-Calél* can be read as an effort to "correct" Zorrilla de San Martín's creationist ideas.

In each of these works, romantic plotlines were the mechanism for grappling with scientifically inspired racial theories and their significance for the nation. Zeballos's fictional works dramatize his scientific theories, and a comparison of Holmberg and Zorrilla de San Martín's poems highlights two very different approaches to understanding racial change over time. At the same time, the literary elements of Zeballos, Holmberg, and Zorrilla de San Martín's works reflect back on the scientific theories they brought to life, revealing the preoccupations with sex and gender that underscored them. In these fictional texts, gender, sex, race, and narrative collide creating powerful depictions of national identity that continue to have impact today. While I do point to some possible sources, I largely follow Gillian Beer's mandate to consider science in fiction through the lens of "interchange rather than origins and transformation rather than translation."[6] Zeballos, Holmberg, and Zorrilla de San Martín's works are not scientific texts and clearly respond to the demands of their fictional genres, but the ways that they draw from, transform, and/or challenge circulating scientific discourses is illuminating.

Prohibited Relations in Estanislao Zeballos's *Painé* and *Relmú*

Zeballos is best known for the nonfictional treatise *La conquista de quince mil leguas*, which guided Roca's Conquest of the Desert, but between 1884 and 1888 he also published a trilogy of texts about life on the frontier in the 1830s. The first book, *Callvucurá y la dinastía de los Piedra* (1884), is a historical chronicle of the life and activities of the Mapuche chieftain Callvucurá. The following two books, *Painé y la dinastía de los Zorros* (1886) and *Relmú, reina de los pinares* (1888), introduce a fictional first-person narrator, Liberato Pérez. Liberato is a frontier renegade who takes refuge among the Ranqueles to escape the forces of Juan Manuel de Rosas, the dictator who ruled Buenos Aires Province from 1829 to 1832 and 1835 to 1852. *Painé* and *Relmú* chronicle his experiences in the toldos, including Liberato's work as Painé's scribe.

In *Painé*, Liberato's narration of his experiences is often interrupted by anthropological and historical information recounted in a voice that sounds very similar to Zeballos's.[7] To establish the veracity of Liberato's observations, the text is peppered with footnotes and citations of real documents. The tension between fiction and reality is most clear in a passage in which Zeballos interrupts his narrator's explanations of the Zorro dynasty's genealogy with a footnote reading, "The skeleton of Mariano Rosas [one of the caciques Liberato is describing], was exhumed by General Racedo and forms part of the author's historical collection."[8] The confluence of perspectives and abundance of ethnographic and historical information in the texts mean that although he is ostensibly an uneducated soldier, Liberato acts and speaks as a fictionalized anthropologist.

The novel also revolves around Liberato's love for Panchita, a white captive who had become the chieftain's favorite wife. Many of Liberato's assessments are shaped by his feelings for Panchita, just as the real-life anthropologists' observations were shaded by their concerns about white women's purity and feelings for indigenous women. The novel reaches a climax when Painé dies and his wives are to be sacrificed in accordance with Ranquel tradition. Desperate, Liberato and Panchita escape from the camp into the uncharted pampas, at which point the novel ends.

Relmú continues the narration of their flight, highlighting their struggles to survive in the face of hunger, thirst, and Indian, creole, and animal enemies. Just when they believe they have reached salvation, Panchita is recaptured in an Indian raid. The rest of the novel follows Liberato as he searches for Panchita and engages in skirmishes related to the politics of the time.

When his military duties take him to the Chilean border, he witnesses the marriage of a chieftain to what he believes is a Picunche women named Relmú. Relmú turns to face the crowd, and Liberato discovers that she is actually Panchita, once again wed to a chieftain. Both Panchita and Liberato faint as he is set upon by the Picunches, leaving open their fate and that of their nation. Zeballos never wrote the finale to the series and readers are left to wonder how he planned to resolve the plot.[9]

In his nonfictional works, Zeballos argued for the cultural and physical elimination of most Argentine Indians, as well as the inadvisability of interracial relations. The romantic relationships in his fiction further develop these ideas in a way that is both straightforward and emotional, encouraging the reader to become deeply invested in Zeballos's racial projects. *Painé* is structured around a skewed love triangle between Painé, Liberato, and Panchita; the end of *Relmú* presents another triangle between Liberato, Panchita, and the chieftain Huamanecul. As already discussed in Chapter One, Zeballos depicts his novelized Ranqueles as inferior by portraying the men as consumed by irrational, animal sexuality. The chieftain Painé is principally characterized as lecherous, marked by a sexuality that is boundless and disgusting. Whereas in Zeballos's anthropological texts this animal lust is directed at white women in general, in the novels, the first-person narrator's own romantic intentions mean that the tragic effects of indigenous masculinity are concretely apparent. Indigenous barbarity does not just affect some abstract "people" or "nation," but rather the romantic couple the narration has asked the reader to believe in.

The romantic triangles in the novels also clearly lay out for the reader which racial pairings are acceptable and productive. Liberato depicts the relationship between Painé and Panchita as disgusting, violent, and coercive, even though, as Susana Rotker argues, a psychoanalytic reading of the novel suggests that Panchita might actually enjoy sex with the chieftain.[10] Through the first person narration, Liberato points out the unnaturalness of Panchita and Painé's relationship and teaches the reader how to react emotionally to this injustice: "Sometimes I shook, blinded by rage, when the Cacique's repugnant lip, in a fit of affection, deposited its lascivious, fetid, and slobbery kiss on Panchita's fine, pink lips" (*Painé*, 288). The juxtaposition of their lips illustrates the incorrectness of the relationship, while Liberato makes sure the reader knows that this transgression of boundaries should cause outrage.

In contrast, Liberato presents the relationship he aspires to have with Panchita as pure and right. Although they barely speak, he claims that they have fallen deeply in love. Given the restrictions on their interactions in the camp, their love appears to stem from nothing more than a shared racial identity and experience of captivity. When they try to escape after eight years, their flight begins with a passionate moment: "Our bodies united in an immense embrace and our lips came together instinctively" (343). In contrast to the mismatch of Painé's and Panchita's lips, Panchita's and Liberato's lips meet as if magnets, further confirming that this racial pairing is natural while the relationship between Panchita and indigenous men is not.

Questions of sex and reproduction support this vision of the ideal racial makeup of the nation. Although Panchita must have had sex with Painé during the more than eight years she is captive, she never has a child with him, suggesting that such relations are not fertile and dodging the most negative consequence of interracial sex: mixed race children. Liberato is also able to reach her before she consummates her second marriage to Huamanecul. Although she has still been touched by indigenous men, justifying their destruction, she does not produce the offspring that would most definitively destabilize the nation. *Relmú* ends with uncertainty regarding whether or not Panchita and Liberato would live happily ever after, but the fact that the white Liberato writes the novels ensures at least his survival into the future. By giving names and faces to the different factions on the frontier, *Painé* and *Relmú* make Zeballos's concerns about indigenous men and mestizaje accessible and engaging for the average reader, even while the abundance of footnotes and historical references remind him or her that the threat is (or once was) real.

Zeballos's fiction also meditates on interracial sex in Argentina through other fictional relationships. As noted above, his narrator describes white captive women as unfortunate victims of hideous, lecherous beasts. In contrast, Liberato rarely talks about the sexual experiences of the numerous white men who are living in the camps. Many surely had sex with indigenous women; indeed, the gauchos and renegades Mansilla met in the same region in which the novels are set spoke openly about the draw of indigenous women's sexuality. Nonetheless, the few times Zeballos's narrator speaks of white men's relationships, he insists that they have taken captive creole—not indigenous—women as wives (*Painé*, 235–38). Liberato's preference for Panchita is reiterated over and over throughout the novels' other romantic pairings.

Furthermore, unlike the Indians' actions, Liberato presents these pairings as acts of mercy by the men who "save" the women from being paired with Ranquel spouses. He commends their willingness to sacrifice "their best garments and their arrogant horses to free those victims shattered by the Indians' lust" (229). The different racial context leads to a reinterpretation of the same act: instead of animal rape, the white men's marriage to captives is practically heroic. For example, in *Painé*, Liberato explains that the real-life rebel coronel Manuel Baigorria had three wives, each a white woman who suffered the horrors of captivity (233). Despite Baigorria's devotion, the first was unable to overcome the "deep bitterness of her savage prison" and died in the toldos (233). The second cried so much for her parents that Baigorria eventually helped her escape (237-38). When the third wife, Adriana Bermudez, betrays him, Baigorria asks the cacique if he can pick an indigenous woman to marry in order to make her jealous (269). Although they get married in a Ranquel ceremony, Liberato does not consider her among Baigorria's "three wives." He also does not describe Baigorria's affection for her as he does the coronel's other companions, highlighting instead her political usefulness and the passion that he inspired in her, just like Liberato would later inspire the Ranquel girl Púlquinay (270). Highlighting these white-white pairings also allows Zeballos to maintain the illusion that white men rarely desired indigenous women and certainly did not reproduce with them. He insists that desire ran in one direction only, from indigenous men to white women.

Again, however, other sources complicate the narrative that Zeballos is telling. Manuel Baigorria's memoirs depict a more expansive and racially inclusive love life. In Meinrado Hux's scholarly edition of the text, he identifies at least eleven female companions over Baigorria's lifetime. While there is confusion surrounding their exact identities and numbers, Hux suggests at least four were indigenous.[11] As Baigorria's experiences show, frontier life was profoundly marked by voluntary interracial relationships of every permutation, many of which were explicitly sanctioned by religious and legal authorities.[12] Zeballos's novels do not depict the frontier as it was, but rather as he wished it to be.

Liberato must also navigate his own romantic entanglements in the pampas, averting interracial disaster. While he pines for Panchita, he is cared for by Púlquinay, a sixteen-year-old Araucanian girl. She cooks for him and protects him from intrigue in the camp. She is also madly in love with him, at least according to Liberato himself (the speed and ease with which Zeballos's female characters fall in love with his protagonist is quite astounding). His feelings toward her are more ambivalent. At first, he wants nothing to do with her: "this woman was truly my guardian angel, although my heart felt dead by her side." Later, however, her abnegation and devotion lead him to wonder if his lukewarm feelings "could reach the vibrating note of love" (*Painé*, 171).

The more distant Panchita seems, the more interested Liberato is in the Araucanian girl. When Panchita roundly rejects Liberato and insists that she will report him to Painé if he talks to her again, he feels as if his life is over. He lists the many things he has lost, and laments that "Púlquinay, only Púlquinay, shone before my eyes with the rosy light of love and life." Together, these changes meant that "Her mouth that smelled like a foal no longer seemed too big, nor did her cheekbones seem to stick out too far, nor ugly her sunken eyes, nor her nose short and round like a turnip. Her whole appearance, which held the promise of the face of a horrifying witch later on, interested me, nonetheless, and I desired her with an infinite zeal between my prisoner's anguish" (181–82). When left with no other options, Zeballos seems to suggest civilized men might understandably feel the temptation to engage in unadvisable relations with indigenous women.

Despite this apparent (and half-hearted) softening, the prohibition against racial crossing quickly rears its head: ten pages later, Púlquinay is found dead with a wound on her breast (204). Even the possibility of a voluntary union between Liberato and an indigenous woman is unthinkable, and it takes her murder to stop it. The fact that she is killed by an injury to her "seno"—translated as breast, bosom, or even heart—further points to loving a white man and the transgression of boundaries as the cause of her death. The romance becomes a cautionary tale, and even the memory of Púlquinay is totally erased as she is never again mentioned in the novels. The unequal and ultimately failed relationship between Liberato and Púlquinay dramatically illustrates the concerns about racial mixing and the unviability of the mestizo that Zeballos developed in his more strictly scientific works.

In some ways, *Painé* and its sequel, *Relmú*, are more conciliatory works than Zeballos's earlier texts, presenting the frontier as "picturesque" and a place of coexistence.[13] Liberato is impressed by the cleanliness of some tribes and finds pleasurable elements in his time among the toldos. At the same time, the treatment of sex and gender make clear that Zeballos's basic project has not changed; in fact, the fictional aspects of the novel intensify it. Indigenous men are inferior men who treat indigenous women as slaves and white women as booty. Indigenous women, while occasionally attractive, are destined to grow ugly. Consensual interracial sex is completely elided. At worst, such relations are presented as coercive and assault, at best, a desperate action when all other options are exhausted. Furthermore, in Zeballos's vision it is always the indigenous person who lusts after the white person: indigenous men take white women captive because they view them to be more beautiful and otherwise "better," Púlquinay shows slavish devotion to the older Liberato for no particular reason. And although incomplete, the story ends when Indians once again impede the population of the desert with good, white families, retrospectively justifying the military actions Zeballos helped facilitate.[14] Through the romance of fiction, Zeballos is able to dramatically represent the racial wars of the River Plate to readers, insisting repeatedly on the inferiority of indigenous peoples and the horror of interracial relations.

Evolution and Degeneration in Verse

Zeballos found what he needed to popularize his racial projects in prose; both Juan Zorrilla de San Martín and Eduardo L. Holmberg turned to epic poetry to productively incorporate past indigenous peoples into national identity. In 1888, the Uruguayan Zorrilla de San Martín published *Tabaré*, an epic poem depicting Spanish-Charrúa interactions on the banks of the River Plate in the sixteenth century. In time for the Argentine centenary in 1910, the Argentine Holmberg published his epopee of the River Plate past, *Lin-Calél*, set just prior to the end of the Spanish colonial period. In each, sexual relations and the gender of the protagonists dramatize the author's preferred theory of racial change and the composition of their ideal nation. Zorrilla de San Martín's text is a defense of creationism and a meditation on degeneration in which the male mestizo protagonist's tragic end paves the way for a Spanish and Christian Uruguay. In contrast, Holmberg defends Darwinism and a mestizo future through the superiority of his mestiza heroine, the eponymous Lin-Calél.

Zorrilla de San Martín began writing *Tabaré* in 1879 or 1880 and first published it in 1888, while Holmberg wrote *Lin-Calél* beginning in 1885 and finally published it in 1910. The two poems are (very) long, have simultaneously epic and elegiac tones, and are set in the era when Uruguay and Argentina formed part of the unified viceroyalty of Peru (later Río de la Plata). They are titled for their mestizo protagonists and each explores the conflict between creole and indigenous society in the River Plate through war and romance. Both end with deaths. Holmberg also explicitly mentions *Tabaré* as an example of another text with a glossary of indigenous words.[15] At the same time, each man was a loud voice in the debates over Darwinism in the River Plate, writing or speaking publicly on many occasions, including immediately after Darwin's death. While Holmberg was an outspoken and well-informed supporter of evolutionary theory, Zorrilla de San Martín attacked it on religious grounds. Zorrilla de San Martín also lived in Buenos Aires from 1885 to 1887, as he was writing *Tabaré*, and certainly would have encountered Holmberg at various events. Viewed through this context, the many similarities of the poems and Holmberg's direct reference to *Tabaré* in the notes suggest that he wrote *Lin-Calél* at least partially to refute Zorrilla de San Martín's arguments with regard to racial change and the incorporation possibilities of indigenous peoples and the mestizo. It is precisely the different genders of their mestizo protagonists that illuminate their oppositional understandings of nature and their contradictory racial projects.

When Zorrilla de San Martín and Holmberg wrote their poems, indigenous populations in both countries had been mostly subdued. This political-military "success" marked a turning point in River Plate science. Once indigenous populations largely disappeared from national political agendas, River Plate scientists ceased to study them as an anthropological other, connecting them instead to a remote past.[16] This archaeologization of the Indian, which led to the museumification hinted at in the fates of Inacayal and his relatives, did not break the ties between science and the symbolic. Scientists turned their focus to prehistory and mechanisms of racial change, particularly the transformation of prehistoric civilizations into inferior vestiges and their failure to survive the encounter with creole populations. Had they progressed from some even more primitive origin, locked in an ultimately doomed competition for survival? Or had they once shared a divine Biblical origin, but degenerated over time?

As active members of the intellectual elites of their countries, Zorrilla de San Martín and Holmberg were both swept up in the debates over evolution that spread rapidly in the River Plate in the last thirty years of the nineteenth century. William Henry Hudson, an Argentine of English descent, received a copy of Darwin's *On the Origin of Species* from his brother shortly after its publication, making him one of the first readers in the region.[17] In Uruguay, the first recorded discussions of Darwin occurred among the members of the Rural Association, an organization founded in 1871 to promote the interests of the agricultural industry. In both countries, the principal tenets of Darwin's ideas were discussed widely and judged on their scientific, economic, and religious value. Within a few years, Darwinian thought permeated major universities across the River Plate, and almost all members of the intellectual elite were aware of evolution, even if they disagreed with the theory. Additionally, the translation into Spanish and massive printing of both *Origin of Species* and *Descent of Man* made Darwin's ideas accessible to a broader River Plate public.[18] This diffusion meant that "most of the literate public had formed some sort of opinion on the meaning of the new science."[19]

Holmberg was one of the most important propagators of evolutionary thought in the region through both his writing and his position at the Escuela Normal de Profesoras (Normal School for Female Teachers).[20] Like most of his peers, he elaborated synthetic understandings of evolutionary paradigms that drew from a variety of thinkers and adapted those ideas to the local context.[21] While Darwin provided the building blocks, Latin Americans like Holmberg were attracted to the moral sense of purpose Herbert Spencer's worldview provided. Additionally, the theories of Ernst Haeckel and Spencer appeared to confirm that evolution was progressive, forward moving, and that human will could correct biological deficiencies. For members of a society that saw itself still in development and often dismissed as racially inferior, this assurance of progress and the ability to influence nature was fundamental.

In contrast to Holmberg, Zorrilla de San Martín was not a scientist and moved in circles that rejected evolutionary thought for religious reasons while often presenting their concerns as scientific. In an article in *La Tribuna* in 1862, Argentine intellectual José Manuel Estrada began, "I do not ask you to respect the Bible, but respect science and respect history . . . respect at least the scientific teachers, whom, although many centuries later, are asserting the same truths that God told to Moses, and that he taught to mankind."[22] He then drew from a vast array of sources, including the works of European naturalists such as Alexander von Humboldt, Georges Cuvier, George-Louis Leclerc (Comte de Buffon), Johann Friedrich Blumenbach, and Petrus Camper, to argue for the unity of the human species and the impossibility of evolutionary ties between men and apes.[23] In Uruguay, the bishop of Montevideo and Zorrilla de San Martín's good friend, Mariano Soler, was similarly careful to unite science and religion in opposition to evolutionary thought.[24] In his texts, geological and paleontological research, the less eugenesic nature of hybrids, and the lack of fossils of intermediary beings all served as evidence of the absurdity of evolutionary theory.[25] The failure of Comte, Spencer, Haeckel, and Darwin to reconcile science and religion was proof that they were "semi-sages" and "scientific riffraff," not proof that the Bible was wrong.[26]

Darwin's death in 1882 presented the opportunity for Zorrilla de San Martín and Holmberg to opine publicly about the polemical scientist and his theories. Holmberg spoke at the homage to Darwin organized by the Argentine Medical Circle, giving an expansive history of evolutionary thought in which he cited dozens of French-, German-, and English-speaking intellectuals. He presented these scholars as precursors to Darwin, a figure he claimed was able to provide "complete development and definitive resolution" to the problem of the origin and progression of organisms.[27] Holmberg dedicated the final portion of his speech to illustrating evolutionary concepts such as artificial and natural selection, the survival of the fittest, and the struggle for life with specific examples from the Argentine context.[28] Throughout, Holmberg drew from Darwin, Spencer, Haeckel, and Lamarck in order to elaborate a worldview in which all organisms—plants, animals, and humans—were locked in a struggle for limited resources and thus survival. Sentiment, mercy, and justice had no place in this perspective, for Holmberg viewed these struggles as natural law, inevitable and immutable.

Zorrilla de San Martín, who Thomas Glick has called Soler's "lay epigone" in the fight against Darwinism, also used the occasion of Darwin's death to express his opinion of the man and his science in an article published in *El Bien Público* (The public good), the Catholic newspaper he directed.[29] In it, he vilified the scientist as "one of the principal enemies of the human species' divine origin."[30] He wrote that although he respected Darwin's geological work and travel writing, he could not accept the revolutionary ideas of *On the Origin of Species* and *The Descent of Man*. Darwin was a good scientist until he grew ambitious and tried to fill in the gaps in the chain of beings, thus attacking Christian faith. As a result of this grave error, Darwin "fell in the most deplorable manner, crushing human dignity and his own glory with the weight of his errors."[31] In 1909 the great Spanish writer Miguel de Unamuno sent Zorrilla de San Martín a copy of a speech he gave on Darwin, suggesting that Unamuno viewed him as someone who still had an interest in the topic.[32] We do not have Zorrilla de San Martín's response or even know if he sent one.

Zorrilla de San Martín's condemnation of Darwin is notably different from those of River Plate intellectuals such as Estrada and Soler. While they fought Darwinism with scientific data, Zorrilla de San Martín relied on emotion and religious indignation. In fact, he did not even appear to know Darwin's home country, calling him a "North American" several times. Although the Uruguayan poet clearly did not understand the science he was attacking, the great popularity of his works both in the nineteenth century and today means that they must be included in any study of River Plate understandings of human origins and racial change.

Degeneration and the Preservation of the Christian Nation in Zorrilla de San Martín's *Tabaré*

Written a half century after the Charrúa Indians were largely wiped out by Uruguayan government forces, *Tabaré* is an elegiac justification of the disappearance of an entire racial group.[33] Set on the banks of the Uruguay River in the sixteenth century, the eponymous protagonist is a blue-eyed child born of the relationship between an Indian cacique and his Spanish captive. After his mother's death, Tabaré is raised with his father's Charrúa tribe. The events narrated in the poem are triggered when the Spanish attempt to found San Salvador a second time. Tabaré falls in love with Blanca, the younger sister of the Spanish military captain Don Gonzalo, and for this he is expulsed from the town. The forest no longer welcomes him either and he wanders feverishly until fainting on his mother's grave. Later, the Charrúas raid San Salvador and Blanca is taken captive by the violent cacique Yamandú. In the end, Tabaré kills Yamandú with his bare hands, rescues Blanca, and returns her to her brother, who believes Tabaré to be the kidnapper and kills him.

Zorrilla de San Martín wrote *Tabaré* in the same period in which he was publishing his anti-Darwinian texts, and many verses of the poem suggest the greater context of contemporary debates over the origin of mankind, racial classifications, and the desire to understand the mechanisms behind change over time. In the opening canto, the poetic voice plants the question that structures the entire text:

> What was that race that passed without a trace?
> Was it the last vestige
> Of a world in decline?
> Twilight without day? Night perhaps
> That emerged dark from the eternal light? (260–64)

Later cantos repeat these doubts:

> Heroes without redemption and without history
> Without tombs and without tears!
> You fought indomitably . . . what have you been?
> Heroes or jaguars? Thought or fury? (1035–38)

Although poetically expressed, behind these lines beat the scientific debates that preoccupied River Plate intellectuals. Where, when, and how did humankind begin? Were indigenous peoples another race, or were all human varieties a single species? How to explain their perceived inferiority: had they not progressed as far, or had they moved backward, degenerating into savagery?

In *Tabaré*, the poetic voice asserts that all mankind was created by God; indigenous inferiority was not due to an original difference but rather their degeneration over time. Speaking of the Charrúas, the poetic voice insists that "In that race, of an exalted origin / a vestige yet remains" (269–70). This explanation protects the Christian narrative of human origins and the creation of all mankind in God's image while simultaneously justifying contempt for the Charrúas and pointing to the obvious political and social superiority of the Spanish. Degeneration becomes the work-around necessary to simultaneously maintain the notions that all are God's children and that the Charrúas were inferior savages who deserved to be killed by the Spanish.

Zorrilla de San Martín uses metaphor and the characterization of Tabaré to appealingly develop this thesis. He connects Christianity and reason to light, tying together two characteristics used during the Conquest and colonial era to justify the oppression of the American natives.[34] Whereas the Charrúas once had that light in their eyes, symbolizing their shared divine origin, it "is being extinguished," marking their degeneration (275). They are "Night perhaps / That emerged dark from the eternal light" (263–64). This light is tied to that which makes people human and keeps them alive, so the poetic voice claims that they are "Twilight without day," and predicts that the Charrúas "will be extinguished" with that light (263, 276). While highly metaphorical, Zorrilla de San Martín's words evoke Robert Knox's 1850 description of the American Indians as a racial group "from whom no good could come, from whom nothing could be expected; a race whose vital energies were wound up; expiring: hastening onwards also to ultimate extinction."[35]

Tabaré's changing characterization establishes a parallel between the historical trajectory of an entire race and the personal history of the poetic protagonist. At the beginning of the poem, Tabaré is characterized as a mestizo and the poetic voice highlights the ways in which he is not like the other Indians. He has light skin, blue eyes, and demonstrates greater emotion and sensitivity then his fellow Charrúas. He also lives on the outskirts of the tribal area and spends long periods of time alone, spatially representing the biological and moral difference between them.

As the poem progresses, however, Tabaré's European characteristics recede and he is ever more identified with the Charrúa masses. Indeed, the poetic voice exhorts the reader: "Look at him. He is the pure Indian; the Charrúa with the narrow forehead" (2031–32). In the end, Tabaré dies and is mourned as "quiet forever, like time, / like his race" (4690–91). Thus, the fictional character of Tabaré undergoes a process of degeneration from baptized, blue-eyed mestizo to a pure Indian, just as in Zorrilla de San Martín's worldview the Charrúas of Uruguay devolved from a noble race into the savages of the nineteenth century. The poetic language and character development in the poem create a parallelism whereby the representation of a fictional being closely tracks the real-life historical destiny of an entire people, making such processes visible for even readers with little concept of the scientific debates.

Zorrilla de San Martín's poem also reflects on questions of racial mixing and hybridity. He depicts interracial sex as one sided and unnatural. Echoing Zeballos and contrasting with Mansilla and Lista, indigenous men lust animally for Spanish women, but Spanish men never desire indigenous women, reinforcing an implicit hierarchy in which the Spanish are superior.

Furthermore, European women do not participate voluntarily in relationships with the Charrúas; all couplings are coercive. As in the works of Mansilla and Zeballos, the indigenous male is depicted as a rapist preying on innocent virgins who suffer greatly at his hand. Tabaré's mother, Magdalena, eventually dies of sorrow after being held captive and raped by his father, Caracé. When Blanca is carried off by Yamandú, Don Gonzalo's biggest preoccupation is for her virginity, not her life: "She is my Blanca! My sister! / Do you remember her? Do you see her? She is not by my side / And she is not dead . . . not even dead!" (3655–57). This characterization of Blanca is strengthened in the next canto when she is described as curled up in the fetal position, a "tight knot" (3807) that cannot defend "the Treasure it guards" against the Charrúas' advances (3809–10). Sexual assault by an Indian is depicted as a fate worse than death.

Zorrilla de San Martín further critiques interracial sex in more abstract terms. Like Arthur de Gobineau, who wrote in 1853 of the "natural repugnance" to the crossing of blood, or Robert Knox's assertion that the amalgamation of races was "impossible," Zorrilla de San Martín argues that, while the races might try to mix, they can never really unite due to an "innate repulsion" between them (105–6).[36] In the introductory canto, a series of highly visual lines reflect on inspiration and poetic creation:

> Forms that pass, bright spots,
> Seeds of impossible existences;
> Absurd lives, in eternal searching,
> For bodies they do not find. (75–78)

Given that Tabaré is repeatedly characterized throughout the poem as an "impossible Indian," it is not difficult to make the connection between these creative processes and the biological mixing of matter to create hybrid humans. Furthermore, in prose and verse statements deleted from the final version of *Tabaré*, Zorrilla de San Martín sympathized with the "anguish the naturalist would have to suffer in order to classify that skull that tried to be an impossible race."[37] Through these passages, Zorrilla de San Martín naturalizes the conflict between the races, making the enmity between the Spanish and Indians a consequence of biology and not personal preference.

As a poetic protagonist, Tabaré collects many of the concerns about hybridity that had circulated in Europe and the River Plate in prior decades. His two sides do not mesh harmoniously but instead engage in a never-ending conflict that reproduces on a small scale the racial wars perceived to be taking place around the globe. His internal struggle is described in a manner both poetic and scientific:

> And the fight of one atom with another
> Renews powerfully in his arteries
> And whistles in his ears,
> And clouds his head,
> And flows to his heart, and explodes there,
> And goes through his flesh and his powers. (2073–78)

The use of the term "atom" emphasizes the entrenched biological nature of his problem. While the mechanisms of genetic inheritance would not be understood until the twentieth century, at the time of Zorrilla de San Martín's writing there was a general understanding that the biological material of the mother and father combined to produce the child. Zorrilla de San Martín depicts this alchemy as taking place in the blood, and Tabaré's mixed ancestry means that the atoms do not work together but continually fight for predominance.[38]

Furthermore, hybrids had been variously lambasted as "debilitated and little vivacious" (29), had "defective prolificness" (42), and exhibited little viability (59).[39] These concerns permeate *Tabaré*, where the title character is compared to a dying leaf cut loose from a sick branch (279–80). In the climax of the poem, he is unwelcomed in both the town and the forest and wanders delirious and staggering like a drunk (2622, 2628–29). For Tabaré, the mestizo condition is one of instability, unproductivity, sickness, and ultimately failure and death, echoing late nineteenth-century concerns over racial mixing as a cause of degeneration.[40]

Zorrilla de San Martín's racial projects cannot be separated from gender. Like his scientific counterparts in Argentina, he depicts River Plate savagery as primarily masculine. The Charrúas in the poem are nearly all men. They are also symbolically associated with darkness, which contrasts with the light of civilization and its divine origin. Drawing on the tenets of *marianismo*, or the Hispanic traditional gender norm that associated women with virginity, maternity, and other characteristics of the Virgin Mary, he depicts the Spanish women as the safe keepers of Christianity and whiteness. This association is further reinforced through women's association with light. Blanca's name means "white" or "brilliant" in its historical root, and doña Luz is literally Ms. Light. While not an allusion to light, Tabaré's mother is Magdalena, connecting her to the biblical Mary Magdalene, one of the followers of Jesus. Through these connections, Zorrilla de San Martín lays bare the religious underpinnings of the creole gender norms that also structured the scientific texts.

In the entry for "Tabaré" in the glossary that accompanies the poem, Zorrilla de San Martín himself makes the argument that gender is fundamental to the construction and interpretation of his work. After explaining the philological roots of the word *Tabaré*, he notes that a friend once asked why he did not use a female figure to personify the Charrúas. His response to the suggestion highlights the importance of contrasts: "No; I had to personify it in the figure of a nearly impossible man, like I could have embodied it in a beast yet unclassified by the scientists, and which, despite being a beast, inspired compassion and even love and tenderness in us" (page 323). It also emphasizes again the gendered aspect of race: savagery is irrational, animal masculinity, while women, associated with concepts such as tenderness and care, are inherently already closer to civilization, even if they are indigenous. A female Tabaré could not have represented the true savagery of indigeneity, especially the sexual threat it presented to white women, nor could she have depicted the warring factions of the mestizo experience. In the end, Tabaré is impossible and must die not just because he is a mestizo, but also because he is a man.

A Darwinist Rebuttal: Holmberg's *Lin-Calél*

Eduardo L. Holmberg's *Lin-Calél* tells a similar story of conflict between cultures. Set in the years immediately prior to the declaration of Argentine independence on May 25, 1810, the poem begins by recounting a conference among all the Argentine Indian tribes in which they debate the impending battle with the bloodthirsty Spanish. A witch's prophecy and the wisdom of the elders determine that the only possible path to victory is for the middle-aged chieftain Auca-Lonco to marry Lin-Calél, the beautiful daughter of the cacique of another tribe and a Christian captive. Reukenám, a young warrior, is sent to announce the prophecy, buy Lin-Calél from her father, and escort her home where she will be the queen of the Andean Pehuenche people. On the journey, Reukenám and Lin-Calél fall in love. In Holmberg's poem, the amorous relationship is between a mestiza and an Indian, as opposed to Zorrilla de San Martín's mestizo-Spaniard relationship, but the pairing ends equally tragically as Reukenám commits suicide by jumping off a cliff upon discovering that Lin-Calél believes in the Christian God.

Throughout the text, Holmberg draws on his experiences as a scientist and explorer to document ethnographic details of various Argentine indigenous groups. Unlike other texts that relied on a generic Indian, Holmberg distinguishes between the assorted tribes and their languages (Pampas, Tehuelches, Pehuenches, Araucanos, Ranqueles, etc.). He describes their customs in great detail, including clothing, greetings, and ceremonies such as the *viñatum*, or parliament. The notes section is rife with references to Argentine scholars, such as Moreno, Mansilla, and Juan Bautista Ambrosetti, as well as varied European, North American, and Latin American authors and texts. These readings, Rubén Dellarciprete argues, led Holmberg toward a more tolerant attitude toward interactions between races and cultures.[41]

Unlike Zorrilla de San Martín, Holmberg adopts the perspective of the indigenous tribes and critiques the actions of the Spanish during the conquest and the colonial period. In their various conversations, they speak of the "invading huinca" (13), "cruel invaders / who outrage the homeland with their footstep" (44), and that "everywhere the furor of the huinca / chases the Indian and mercilessly kills him" (47). Whereas Zorrilla de San Martín gave space to the argument for natural slavery voiced by sixteenth-century theologian Juan Ginés de Sepúlveda, Holmberg repeats almost verbatim the arguments of Sepúlveda's adversary Bartolomé de las Casas, including the image of Spaniards killing indigenous nurslings (50).[42] Reukenám also implies that the Spanish will rape Lin-Calél horrifically (144), approximating the Christians to the dehumanized and animal-like behavior of the Charrúas in *Tabaré* or the Ranqueles in Zeballos's novels.

Fernando Operé, one of the few critics to mention *Lin-Calél*, calls it "a posthumous and precursor poem" in relation to indigenista literature. "It closes the chapter on a literature that justified military actions that pushed frontier tribes to an unrecognizable south, and it initiates a new Indianist literature marked by a badly assimilated sense of guilt."[43] While Operé is correct in locating *Lin-Calél* at this inflexion point, Holmberg's scientific proclivities preclude the possibility of guilt in the poem. In his homage to Darwin, Holmberg began to develop his understanding of the relationship between evolution and the Argentine conquest of the pampas. Speaking of the recent actions of the Argentine military, he argued,

> Is the Indian's cause just? Arguing without much dialectic, the Indian defends his land, which we have usurped, wounds us, kills us, steals from us. Does he do good? It is good, or not. He struggles for life. . . . Whites, civilized men, Christians, armed with Remingtons, we do away with the Indi-

> ans, because the Law of MALTHUS is above those individual opinions . . . and so, we too fight for life, with good ideas, with good weapons, with good resources, we do nothing but put our advantages into play.[44]

Rather than highlighting guilt for the Argentine extermination of its indigenous peoples, Holmberg's perspective actually erases any blame that could be placed on the Spanish, and by extension, the Argentines, for this genocide. Malthus's Law, a purely mathematical issue, is the backbone of the struggle, thus removing the question of morality or even human emotion. The Indians are not cruel barbarians for defending their land, nor are the Argentines wrong for using guns and other resources to battle back. For Holmberg, the Indian/Christian conflict was inevitable, and victory would be determined not by divine mandate or moral superiority but rather by the immutable and inescapable laws of Nature. Holmberg thus naturalized the extermination of the Indian as a necessary and amoral step in the processes of Darwinian evolution that would lead eventually to perfection.[45]

Lin-Calél dramatizes this fight. The indigenous tribes and the Spanish fight over land, resources, and women, breathing life into a Malthusian understanding of the world. Interestingly, the voice that rationalizes evolution in Lin-Calél is that of the Indian, not that of the Spanish, as Holmberg repeatedly puts the words of Darwin and Spencer into the mouths of his indigenous protagonists. For example, the warrior Reukenám contextualizes the fight against the cristianos with a Spencerian understanding of the world:

> That is the law of life, the struggle
> To defend the sacred fatherland
> Ground where old champions lie
> Who died fighting to defend it. (145)

Life is a struggle, and the Indians and Spanish must try to exterminate each other in order to gain access to the scarce resource of land. To not fight is essentially to commit suicide. Holmberg reiterates the amorality of the struggle in the notes section of the poem. While admitting that the Indians actually treated the Spanish quite well during the Conquest and reaped no rewards, he insists, "there was a need to kill and teach cruelties" (327).

In *Lin-Calél*, the indigenous protagonists also make pronouncements on the relative fitness of the two warring groups, admitting that they are the weaker combatant and predicting their own defeat. During the parliament to determine a plan of action, the cacique Numillán details the Spaniards' unsurpassable advantage:

> The weapons
> Of the cruel invader, more effective
> Are than ours . . .
> [. . .]
> The stone kills in combat, and the knife
> In a skilled hand is as useful as a sword;
> But all the warriors' courage
> Is sterilized when bullets whistle. (49)

According to him, the Spanish had evolved further than the Indian in terms of culture and innovation, developing weapons against which the native tribes could not compete. Evoking Lubbock's progressive classification of societies according to the materials of which their tools were made, Numillán insists that the indigenous groups still straddling the Stone and Iron Ages cannot obtain victory over the technologically superior Spanish. The struggle between the Spanish and the Indians is thus a battle between different time periods, again excluding any question of morality or justice. By putting the evolutionary justification for the extermination of indigenous peoples in their mouths, Holmberg further naturalizes the process and thus the outcome. If even uneducated Indians intuitively sensed the mechanisms of evolution, then the theory must be a valid explanation of the historical moments that shaped the physical and human landscapes of Argentina.

Holmberg's understanding of life as a struggle for survival means that, once again, indigenous peoples are destined to vanish due to their inferior fitness. As in *Tabaré*, in *Lin-Calél* the male protagonist is emblematic of the entire native race and its fate. In *Lin-Calél*, however, this synecdochic character is of pure Indian blood. Whereas eighteenth- and early nineteenth-century conceptions of progressive theory, such as those of Lyell and Lamarck, maintained that all human groups could, with proper intervention, reach the upper levels of civilization, the work of Darwin, Spencer, and others in their vein injected competition and thus violence into the processes of progress. As Adriana Novoa and Alex Levine note, "By the end of the century, the progress of civilization had come to require the replacement of the individual *physical* bodies that comprised the nation. Its evolution must now be conceived, in Darwinian terms, as entailing *waste*."[46] Savage peoples could no longer be expected to evolve past a certain point for their relative lack of fitness would dictate the truncation of their line at an earlier stage; Reukenám's body flying over the abyss symbolizes the waste of Darwinian extinction. Despite their contradictory methods of racial change, in both *Tabaré* and *Lin-Calél* the pure-blooded Indian (always depicted as male) is doomed to disappear. He is either necessary waste in an evolutionary process or a ruined degenerate too far from divine origins to survive.

Hybridity and the Nation in *Tabaré* and *Lin-Calél*

Although both *Tabaré* and *Lin-Calél* end with death and destruction, they do not espouse a nihilistic view of nothingness in the River Plate. Instead, they foreground the disappearance of the native as a necessary condition for the coming of a new race. Both poems speak of a "raza nueva" (new race) that was about to arrive, and both explicitly tie this new race to their modern nations (*Tabaré*, line 360; *Lin-Calél*, page 276). Gender again provides the key for reading these endings, as the different genders of the protagonists determined their fates and thus their symbolic relationship to the new race and the nation in formation.

In *Tabaré*, the pull of savage masculinity motivates Tabaré's degeneration over the course of the poem from an association with mestizaje to a purely Indian characterization. His death in the end clearly indicates that the mestizo race was as unwelcome in Zorrilla de San Martín's future Uruguayan nation as the Charrúas were. Zorrilla de San Martín emphasizes this point in the dedicatory letter to his wife that precedes the poem. He writes that Blanca is part of "your race, our race" and that "the last notes that you will hear in my poem are the laments of the Spanish woman and the monk's prayer, the voice of our race and the accent of our faith" (page 92). A direct connection is established between the sixteenth-century Spanish arrivals to the Americas and the nineteenth-century Creoles. For Zorrilla de San Martín, Uruguay is a Spanish and Catholic nation and could not have been anything else. In other texts, he reaffirmed this racial makeup by connecting Uruguay to the "Spanish family" and "the Caucasian [race], which is ours."[47] Over and over, Zorrilla de San Martín denies the Indian and mestizo heritage of the continent and affirms Spanish influence.

In the end, the Spanish Blanca is left unsullied so that in the future she can reproduce to populate the nation. She is therefore Uruguay's literal and symbolic womb. Women were frequently cast in this role by nineteenth-century anthropologists such as the North American Daniel Garrison Brinton, who wrote that white women had "no holier duty, no more sacred mission, than that of transmitting in its integrity the heritage of ethnic endowment gained by the race through thousands of generations of struggle."[48] This idea repeated itself in Spanish American literature of the era, the most effective example of which is Esteban Echeverría's *La Cautiva*

(1837), a similarly long poem about the desert, Indians, and national identity. Echeverría constructs the female protagonist María as a symbol of the Argentine nation. Her greatest accomplishment in the poem is remaining untouched by the Indians, thus ensuring the purity of the national body.

Likewise, in *Tabaré*, Blanca's value is conceived of almost entirely in reproductive terms. Her virginity is repeatedly highlighted, and she is often described as a "niña" or child so as to further distance her from the passion exhibited by the Indian chieftains. In the third canto of the second book, a parallel is established between Blanca, the Virgin Mary, and Tabaré's mother, highlighting Blanca's dual characterization as both virginal and maternal (1796–99). Although currently uninitiated into sexual relations, she already possesses the characteristics necessary to birth and care for the new nation. This is the reason it is so important she not be raped by the Indians.

Tabaré promotes a providential and teleological vision of the history of the Uruguayan nation that leads to a pure, unadulterated populace still firmly connected to the birthplace of the "noble mother race" (805), Christian Spain. The project depicted in *Tabaré* was clearly articulated by Zorrilla de San Martín six years later in a speech at the Ateneo in Madrid on January 25, 1892:

> But I am convinced that this man was not and could not be a beginning; he was an end, a last vestige. Nature was young and beautiful, and man was decrepit; he was near death, and nature was being born or reborn; man feared and noted ominous omens everywhere, and nature worried; man dug his grave, while nature covered that grave with moss and flowers, and prepared in it a cradle or a marriage bed for the man she was waiting for or sensed, able to understand her, to love her, and to make her a mother.[49]

Here, Zorrilla de San Martín makes explicit the separate trajectories of the races living in sixteenth-century America. The images of the tomb and the womb overlap, demonstrating that as the Indian came to the natural end of his line, the modern Uruguayan nation was just beginning to flourish with the divinely ordained support of the feminized American land. Thus, in Zorrilla de San Martín's vision, the modern Uruguayan nation is composed of the White, Christian race that God created to fulfill the destiny of the Americas after the indigenous tribes were lost to the unforgiving forces of degeneration.

Lin-Calél also mirrors the processes of racial change with those of nation formation, carefully tying the struggles for survival to the necessary struggles to create a cohesive, productive nation. The winner of the evolutionary battle in the poem is the future Argentine nation. As in *Tabaré*, tomb and womb go hand in hand as the struggle between contemporaries is recast as a struggle between past and future. Parallel to the premonitions of their own death, the Indians in *Lin-Calél* have visions of "the race that will come with new vigor to proclaim the freedom of America" (276). Unsurprisingly, this new race is associated with the Argentine nation through the image of "something blue and white in the mountains," a reference to the sky blue and white national flag. This connection is made even more obvious by the fact that the last canto is set on May 25, 1810—the date of the foundation of the first national government—and the fact that Holmberg published the poem just in time for the 1910 centenary of that moment. In Holmberg's vision of the world, highly influenced by Darwinism, the struggles of the conquest and colonization of Argentina could be viewed as the playing out of the natural and biological laws of the world. The Indian, biologically inferior to the Spanish, would lose the battle and give way to a superior new race.

Unlike *Tabaré*'s entirely white, Spanish, new race that maintains only a poetic memory of the Charrúas, Holmberg insists that the mestizo race will found the new nation. In his poem, it is the pure-blooded Reukenám who dies, while the mestiza Lin-Calél is left as the symbolic representative of the future. Like Blanca, she is a woman and thus the womb that will reproduce the new nation. Holmberg writes this explicitly in the glossary: "[Reukenám] is the incarnation of the Indian soul, as Lin-Calél is of the New Race" (347). In order to prepare her for this outcome, she is associated with maternal qualities throughout the poem. She ministers to the sick, lovingly calls the children, and even milks a cow and feeds the milk to the children in a strange, somewhat sexualized allusion to breastfeeding (168). She is also virginal, making her an ideal vessel for the new nation (176). Although their scientific and religious convictions were quite different, Holmberg and Zorrilla de San Martín appear to agree on the ideal characteristics of a woman: pure, maternal, and closely connected to spirituality.

Accordingly, Holmberg's vision of the mestizo (mestiza here) is much more positive than the negative Tabaré. Lin-Calél is exalted and passionately pursued by a number of Indians. She is so beautiful that there is no one anywhere that compares. Rumors swirl that many had died for her love. She is also clean, orderly, useful (she can cure the sick, knit, weave, cook, paint, and even read), and very smart (150). The very qualities that most attract the Indians are the ones associated with Lin-Calél's captive Christian mother, who educated her before being thrown out of the toldos for praying in the Christian fashion. Lin-Calél's racial otherness makes her an object of desire while the blind devotion of the Indians reaffirms the superiority of the Spanish, permanently inscribing them in an inferior position lusting after an unattainable superior woman.

Nonetheless, Lin-Calél's exceptionalism does not come entirely from her Spanish side. In many ways, she is also the pinnacle of positive indigenous qualities. She is able to ride like a man and is strong and vigorous. Despite Reukenám's suggestions, she emphatically insists that she does not need to stop and rest during their long journey, nor does she need help to ride a horse (234). Her father explains to Reukenám that Lin-Calél is truly unique: "There are not women like her among the Christians, / so vivacious and so skillful, so active" (129). Whereas a Spanish woman may also have been able to read or care for children, Lin-Calél possesses energy unknown in white society. Holmberg's characterization of Lin-Calél thus supports the theory of hybrid vigor, which argued, as James Cowles Prichard summarized, that "mixed descendants are physically a more vigorous offspring."[50] In Lin-Calél, the mixing of two races was a successful experiment, giving rise to a being superior to either of her two lines of ancestry and directly contradicting the dire predictions of mestizo monstrosity literarily represented by Zorrilla de San Martín in *Tabaré* and by Zeballos in his novels.

Lin-Calél's great beauty and ability also allow Holmberg to incorporate sexual selection into a poem that had thus far been dominated by concepts of natural selection and its corresponding extinction. Like many Latin American intellectuals, Holmberg was very interested in sexual selection because it presented the opportunity for humans to influence evolution and thus the future of the nation.[51] Although barbaric, Tromen-Curá's capture and rape of Lin-Calél's mother because of her Christian beauty does lead to a bettering of the race in the form of Lin-Calél. Additionally, the *machi* (witch doctor) predicted the marriage of Lin-Calél to the vile chieftain Auca-Lonco, whom Lin-Calél finds horrific. She instead choses to flee with Reukenám, a young and gentler warrior, demonstrating Haeckel's idea of female selection. The machi's prophesy, like the biological determinism inherent in natural selection, is overruled by human action, altering the course of development and ensuring a more positive outcome.

Additionally, while natural selection could be used to justify the disappearance of indigenous peoples, sexual selection was a way for Holmberg to preserve elements of the less fit race. Holmberg's indigenous protagonist commits suicide, symbolizing the physical extinction of his race. The temporal setting of the poem further relegates him to prehistory. The fact that he dies on the day Argentine independence is declared suggests he is a precursor to the nation but will not actually be part of the state. But Reukenám's death is not the total disappearance of indigenous qualities. When he jumped, he left behind

> A seed of courage and vigor,
> A type in gestation which in its softness
> Will become the masses,
> And docile to the model offered to it,
> Shall sink into the bosom of libertines,
> Or rise to the summits of Glory
> On the winds of atavistic virtue.
> Here's to you, seed of the uncertain future! (306–8)

The language in these stanzas is both highly symbolic and very scientific, as the exalted cries of patriotism and phrases such as "the summits of Glory" contrast with terms such as *gestation* that only entered the language in the nineteenth century through science.[52]

Although we have no evidence that Lin-Calél and Reukenám had any sort of physical relationship, the use of "seed" and "type in gestation" in the last verses makes clear the connection between the symbolic uniting of the races described in the text and the actual sexual contact that had to occur in order to produce Lin-Calél and others like her. In Holmberg's vision, this contact would be produced again and again in the formation of the Argentine people. The mestizo is not only a necessary part of the nation but is also a productive and fertile source of future citizens.

With its eclectic piecing together of evolutionary theory, this ending coincided with general attitudes in the River Plate. For relatively new nations dealing with varied racial populations, the pessimistic and even chaotic vision of the world in Darwin's work was politically unviable and terrifying, thus scholars gravitated toward more reassuring progressive explanations.[53] *Lin-Calél* demonstrates this desire, for despite Holmberg's insistence that he is a committed follower of Darwin, the past is clearly depicted as leading toward a particular point. Although the mechanism is different from in Zorrilla de San Martín's poem, Holmberg's text still maintains a version of history that—like that of his preferred evolutionary thinker, Haeckel—emphasizes forward motion toward unity and perfection: the new race is "a Sun of glory / a Sun of freedom, bright and new" (281). Due to the needs of the nation and the moment when the poem was published, Holmberg's Darwinian assertions are tempered by the overall feeling of a teleological history leading to a particular and superior outcome.

Lin-Calél also reveals a continued reliance on Lamarckian soft inheritance and Haeckel's "progressive heredity," or the transmission of acquired characteristics. In these theories, learned traits could be passed on to future generations, allowing human agency and accumulated knowledge to raise the level of entire races. The seed that Reukenám leaves behind will be "docile to the model they offer it," succeeding or failing due to education and experience. Despite the fact that the work of August Weismann and Gregor Mendel was fairly well-known by 1910, these verses indicate that for Holmberg, the result of the mixed-race experiment was not totally determined by its genetic components.[54] The words "uncertain future" further suggest the limits of biological determinism. The ending of the poem depicts an evolutionary world in which concrete actions—education and the hygienic practices that would come to dominate the early twentieth-century River Plate experience of immigration—could shape the genetic heritage of the nation. Once again, these modifications to Darwinian theory created a world in which Argentine elites could continue to alter the biological materials they were given in order to create the ideal nation. The triumphant ending of Holmberg's poem popularizes this scientific thinking while arguing that this ideal would rely on biological input from indigenous and nonindigenous actors.

Scientific Accuracy, Aesthetics, and the Canon

Fiction is not a field usually associated with scientific thinking, but for Lista, Zeballos, Holmberg, and Zorrilla de San Martín it was a fertile space for exploring the latest theories of race and what they might mean for their nations. By fictionalizing creole-indigenous conflicts in the River Plate, they were able to investigate national origins, change over time, and possibilities for the future. The gendered elements of romance made obvious the fact that, as Robert Young insists, nineteenth-century racial theories were fundamentally about heterosexual sex.[55] These texts were also a way for River Plate intellectuals to continue popularizing their theories, as the emotion and literary devices of the texts made their ideas more accessible for a general audience. Without fiction, we cannot understand how elite scientific projects become collective identities.

Fiction also shows why identities last even when the science behind them has been disproven. Comparing *Lin-Calél* and *Tabaré*'s scientific premises and interpretations of River Plate history, it is clear that Holmberg's vision, while quite imperfect, has held up better over the years. Degeneration as a mechanism of racial change has been discredited and evolutionary explanations continue to thrive. Furthermore, recent genetic studies of the Argentine and Uruguayan populations have demonstrated the continued presence of indigenous genetic material, more closely approximating Holmberg's "raza nueva" than Zorrilla de San Martín's. In Argentina, investigators at the University of Buenos Aires found that over 50 percent of the population had indigenous ancestry on mitochondrial DNA analysis.[56] In Uruguay, that number is 31 percent.[57] Interestingly, the same studies show that the percentage of Native American ancestry on mitochondrial DNA tests, which measure the maternal line, is many times higher than the percentages found in Y-chromosome (paternal) DNA testing. This difference demonstrates that the racial relations the anthropologists fretted over—indigenous men with white women—were far less common then the relationships between white men and indigenous women that Zeballos was so eager to belie.

Scientific accuracy does not guarantee the popularity of a literary text, however, and thus is uncorrelated with the lasting effects of its images. Despite Holmberg's more accurate understanding of race and reproduction, his poem has quickly fallen out of favor. Indeed, only one thousand copies were printed, despite an excerpt published in the popular weekly magazine *Caras y Caretas* to increase interest.[58] Dellarciprete argues the poem was not well-received because its vision of mestizaje was out of step with that promoted by Centenary elites including Leopoldo Lugones, Manuel Gálvez, and José Ingenieros.[59] It is also rather stilted and unexciting. In contrast, *Tabaré* was well-received by both the Uruguayan elite and the general public and quickly reached canonical status. Since its publication, *Tabaré* been translated into French, Italian, English, German, and Portuguese and reinterpreted in a number of mediums, including opera, painting, and sculpture.

Although not scientific, for many Uruguayan readers in the nineteenth century and today, *Tabaré* shaped their understandings of indigenous peoples, how humankind changes over time, and Uruguayan national identity. The literary power of *Tabaré* and its romantic entanglements have ensured that Zorrilla de San Martín's vision of the Indian and racial mixture have lived on. The differing fates of Zorrilla de San Martín and Holmberg's poems show that even when proven false, the romance of fictional approaches to nineteenth-century scientific thinking could be even more influential than the empirical texts from which they drew, shaping national identities at the time of their publication and centuries into the future.

CHAPTER 5

Defiant Captives and Warrior Queens

Women Repurpose Scientific Racism

Most Argentine anthropologists presented natural history as a man's world, employing the tropes of the penetration of barren Argentine wilderness and the fertilizing powers of civilization. Woman, when mentioned, was the object of study, never the observer. Nonetheless, in 1860 an upper-class Argentine named Eduarda Mansilla published a short novel, *Lucía Miranda*, that included ethnographic-style descriptions of the Timbú and Charrúa tribes. Two decades later, she travelled to the United States and condemned the Native American genocide there in her 1882 *Recuerdos de viaje*.[1] Around the same time, Scotswoman Florence Dixie left behind her two children—one a newborn—and set off for Patagonia with her husband and two brothers. They spent several months hunting, meeting indigenous peoples, and enjoying the splendor of riding across virgin territory; Dixie recorded her observations in a popular travelogue, *Across Patagonia* (1880).[2] These experiences would later inform two children's novels, *The Two Castaways* (1890) and *Aniwee, or, The Warrior Queen* (1890), that also dialogue extensively with George Chaworth Musters's *At Home with the Patagonians* (1871).[3] Mansilla and Dixie's texts demonstrate that many elite white women were not only aware of many of the tenets of scientific racism and its consequences for colonizing projects but also appropriated those discourses to challenge exclusionary masculine actions and advocate for more opportunities for both women and indigenous peoples.

Argentine women did not formally participate in the natural sciences until the early twentieth century.[4] There are examples of earlier female activities, but seeing them requires what Fontes de Oliveira calls "A reconsideration of premises about what constitutes science and a reevaluation of its gendered paradigms."[5] In the late nineteenth-century, several women worked in the Museo de La Plata's publications department as "bookbinders and paper folders" or cared for animals on the premises.[6] Others corresponded with and/or provided materials for the Museo de La Plata and Museo Etnográfico right around the turn of the century.[7]

Additionally, Francine Masiello and Bonnie Frederick have both made the argument that women engaged with conversations about race, nation, and progress as they worked to identify their own position in Argentine society. Both authors associate the nineteenth-century family with the emerging nation and claim that within this unit, women were seen as forces of stability and civilization, responsible for the education of future citizens.[8] This framework intimately connected the fight for women's rights with the *cuestión de indios* by tying the "triumph of the European population" in the Conquest of the Desert to women's progress.[9] These connections show Argentine women had access to the major debates of the time, an assertion confirmed by newspaper reports on scientific speeches and events, particularly those held in Buenos Aires. For example, *El Nacional* noted the presence of many women at the homage to Darwin during which Holmberg and Sarmiento spoke.[10]

Eduarda Mansilla and Florence Dixie are two examples of female writers who, although of different nationalities, shared an upper-class identity, international perspective, genre-spanning professional writing career, and interest in using the Argentine frontier to articulate projects of racialized and gendered identity. Both were outsiders to the regions and cultures they represent. Dixie was a Scotswoman in Argentina; Mansilla visited the US and, in her writing, time-traveled to an Argentina where indigenous cultures still flourished, unlike the Buenos Aires in which she lived. The fact that both women used indigenous peoples to develop their texts further justifies their joint study. While other late nineteenth-century Argentine women writers, including Lola Larrosa, Josefa Pelliza de Sagasta, María Eugenia Echenique, and Raymunda Torres y Quiroga used the language of progress and evolution to argue certain rights and positions for women, Eduarda Mansilla was one of few to write explicitly about indigenous peoples as the background to these debates.[11] In this sense, her work can be explicitly compared to Dixie, who develops an entire indigenous world in her children's novels.

More importantly, both Mansilla and Dixie develop *female* indigenous protagonists. As I have argued throughout this book, Argentine men relied heavily on physical and literary indigenous women as they developed their scientific-political projects. Through Mansilla and Dixie, we can see how women used similar literary figures to carve out spaces in the debates over national identity and colonialism to examine their own subaltern position, as well as how these positions and their nationalities led to alternative assessments of indigenous peoples' potential. Although not scientists themselves, both women were very connected to the scientific worlds of their respective nations and their commentary on colonial projects—although only indirectly dialoguing with racial science—is likely the closest we can get to knowing women's perspectives on the themes that have structured this book.

Mansilla was raised in a wealthy family that valued education, and her brother, Lucio, wrote one of the classic texts of nineteenth-century Argentine ethnography. Her journalism appeared in the major Argentine newspapers where scientific debates played out, including *La Tribuna*, *El Nacional*, and *La Nación*. In her articles, she mentions reading Auguste Comte and twice references Darwin.[12] Through her husband, her brother, and her own outstanding participation in elite cultural circles, she also had privileged access to the men making science and policy decisions in Argentina at the time. As Fontes de Oliveira claims, "Mansilla may not have followed natural history paradigms, but she was well aware of dominant discourses that name, classify, and categorize nature and people to best fit ideological interests."[13]

Like Jose Manuel Estrada, Mansilla was generally critical of positivism and evolutionary theory, repeatedly claiming to support science until it impinged on Catholic faith. For example, in her article on women's education, Mansilla wrote, "Science is a great thing! Faith is even more so!"[14] She continued by criticizing Sarmiento's Protestant North American teachers, insisting that one could be educated and Catholic, know and teach the latest human knowledge without clashing with "the consoling beliefs of our daughters."[15]

Florence Dixie's scientific aspirations and ties to Argentine and British naturalists were more explicit. Although critics frequently classify her as a tourist traveling for pleasure, Dixie's travelogue, *Across Patagonia*, adopts and winkingly subverts many of the conventions of scientific travel writing.[16] She corresponded with Charles Darwin, sent him a copy of *Across Patagonia*, and corrected his description of the *tuco-tuco* rodent. Furthermore, *Across Patagonia* and Dixie's two children's novels clearly dialogue with George Chaworth Musters's *At Home with the Patagonians* as Dixie borrows names, indigenous words, theories, and even entire passages.[17] Although the two women's texts were not intended to be scientific, it is clear they were aware of various contemporary scientific ideas and that these, including the representations of indigenous peoples and nonindigenous Argentines produced by masculine science, gave shape to their writing. By looking at Mansilla and Dixie's texts in this context, we can continue Chapter Four's examination of the colorful life that scientific ideas take on outside the formal bounds of science as they are transformed by nonscientist authors, the demands of genre, and competing political projects. Dixie and Mansilla also allow a return to the discussion in Chapter One of the ways that elite women reaffirmed or pushed back against hegemonic projects of identity and colonialism.

Comparing Mansilla and Dixie is also an effort to bridge a theoretically based gap in the literature. Although there are many similarities between Mansilla and Dixie, only two critical texts unite them. The first is Mónica Szurmuk's study of female travelers in Argentina, although she does not compare Dixie and Mansilla so much as allow them to exist side-by-side. The second is Natália Fontes de Oliveira's book on Dixie, Mansilla, and Elizabeth Cabot-Agassiz. This disconnect is emblematic of a larger challenge in the literature on both colonialism and history of science. Argentina and many other Latin American countries shared characteristics with colonial and imperial contexts around the world. Nonetheless, there is surprisingly little dialogue between historians and scholars of cultural studies who focus on the English-speaking world and those who specialize in Latin America. For example, Syed and Ali's 2011 analysis of the white woman's burden is divided into sections labeled "South Asian," "African," "Australian," and "American" contexts. Despite the fact that Argentina was a settler colony like Australia and the United States, the "American Context" focuses entirely on the United States, excluding Spanish-speaking America.[18]

Most histories of anthropology also focus on the English-speaking world, and there are a scant few texts that compare Latin American science to that happening in the United States, Canada, or other settler colonies. Similarly, critics focused on Latin America rarely situate the texts they study within global trends, taking nationalistic or regional approaches to historical phenomena. This gap may be explained by language issues as well as the fact that Latin America is generally not considered "postcolonial." In fact, many Latin American intellectuals have been skeptical of the applicability of postcolonial theory to the region.[19] In her book, Fontes de Oliveira uses the examples of Dixie, Mansilla, and Cabot to argue for the fruitfulness of studying English-language and Latin American writers together.[20] I agree and add that we must go beyond disciplinary boundaries to study nonfiction and fictional works together. This chapter thus aims to highlight the parallels between British and Argentine projects, as well as the differences seen in Mansilla and Dixie's work.

Such an analysis is especially appropriate as Argentine frontiers and identities had been forged in a context of Anglo-Latin contact. From the 1806–1807 invasion of Buenos Aires to mid-century attempts to attract immigrants to the influx of ranchers and missionaries, the nineteenth century was marked by British-Argentine political and economic entanglements. As we have seen at various points in this text, mid- to late century racial science trafficked in Anglo-Latin comparison, often negatively juxtaposing the Latin South American republics to the United States. While the United States was obviously no longer part of Britain by the late 1800s, many of the discussions of United States progress were explicitly or implicitly grounded in its Anglo-Saxon racial origins. Both Dixie and Mansilla engage with this background: Dixie criticizes Argentine treatment of indigenous peoples to justify British expansion, while Mansilla compares the Latin and Saxon races as well as Argentine and US Indian policy. In doing so, she frequently stresses the United States' British political and racial inheritance (*Recuerdos de viaje*, 78). Comparing their texts thus makes clear the ongoing physical and ideological tensions between the Anglo and Latin worlds.

Mothering the Children of the Desert: Eduarda Mansilla's *Lucía Miranda*

Today, Eduarda Mansilla is renowned as one of the first female Argentine writers and her works are the subject of an ever-growing body of critical literature. Born in Buenos Aires at the end of 1834, she belonged to an aristocratic family with close ties to Argentine political and cultural elites. Her uncle was Juan Manuel de Rosas, the iron-fisted governor of Buenos Aires who ruled from 1829–1832 and 1835–1852. Other relatives, like her brother Lucio, played public roles as military men, politicians, and authors. In 1855, Eduarda married Manuel Rafael García Aguirre, a diplomat with whom she would have six children. Together, they made multiple journeys to Europe and the United States.[21] In 1860, Mansilla published her first two literary works, *El médico de San Luis* and *Lucía Miranda*, under the pseudonym Daniel. A novel in French, *Pablo, ou, La vie dans les pampas*, an account of her travels, several plays, and multiple collections of short stories followed. Mansilla also made a name for herself as a journalist and composer, and her intellect earned her the admiration of personages such as Abraham Lincoln, Henry Wadsworth Longfellow, Victor Hugo, Alexandre Dumas, and numerous European royals.

Lucía Miranda, one of Mansilla's first published works, draws on the legend of a Spanish woman, Lucía, who was taken captive in Argentina around 1532. Although not based on any specific historical event, the tale has featured in multiple Argentine and international texts since its first appearance in Ruy Díaz de Guzmán's *La Argentina manuscrita* (1612).[22] The nineteenth century saw a proliferation of Lucías in prose, verse, and drama; Mansilla published *Lucía Miranda* the same year that Rosa Guerra published her novel of the same name.

The first half of Mansilla's version is dedicated to depicting Lucía's lineage, her early education, and the lives of the people around her. These include her adoptive father, the soldier don Nuño de Lara; a kindly priest who encourages her to read, Fray Pablo; and her future husband, Sebastián. Throughout this backstory, accounts of orphaned children adopted by others abound, as do frustrated loves and reminders of the importance of women's education. The second half of the novel begins when Lucía, Sebastián, Fray Pablo, and don Nuño cross the Atlantic, settling at the Espíritu Santo fort at the confluence of the Carcarañal and Paraná rivers (145, 147).[23] Peaceful relations with the kindly Timbú Indians mark their first few months. Lucía increasingly acts as a spiritual and moral guide for the indigenous women and serves as cultural mediator for the Spanish authorities. She also encourages a relationship between Alejo, a Spanish soldier, and Anté, an indigenous woman. Before long, however, the cacique Marangoré falls in love with Lucía's goodness and beauty. His jealous younger twin, Siripo, encourages him to break his promises, attack the Spanish fort, and kidnap Lucía. The Spaniards are massacred and Siripo murders Marangoré, taking Lucía for himself. She refuses his advances, choosing death over voluntarily giving in to the Indian's lust. In the end, both Lucía and Sebastián are thrown on the pyre as Alejo and Anté flee, seeking shelter in the pampas.

Critics have widely seen Mansilla's colonial setting as intimately connected to her vision of Argentina's future, but there is little consensus on the specifics of that vision. Mansilla published the novel in the middle of a period of reconceptualizing national and group identities as Buenos Aires and the Confederacy jockeyed for position. Although nearly twenty years before the Conquest of the Desert took "final" action against the indigenous tribes of the interior, the question of indigenous people's presence in the nation's landscape and genotype was of great interest to intellectuals. Frederick, Masiello, and Rotker have argued that Mansilla uses the colonial legend to reinforce the Indian's savagery and justify his or her exclusion in the present, while Barraza Toledo claims that Mansilla's "feminine project of pacific conquest" fails, throwing into doubt the worthiness of such a project.[24] Hanway takes a middle ground, arguing that Mansilla allows mestizaje but only in the pampas, "an extra-legal space that brings love and trouble in equal quantities."[25] In contrast, Pérez Gras insists that Mansilla is pro-mestizaje and that the violence of Marangoré and Siripo is mere plot device.[26] Despite these proclamations, few of these texts devote more than a few paragraphs to race.

With its publication in 1860, Mansilla's *Lucía Miranda* antedated the peak of Argentine anthropology, including the publication of each of the texts studied in this book. Nonetheless, it shares many literary and historical roots with them, including Sarmiento's *Facundo*, the scientific-historical travelogues of preceding centuries, and the legacy of Mansilla's uncle's campaigns against the Indians of the desert in the 1830s. Furthermore, as in the male anthropologists' fiction, in *Lucía Miranda*, national identity and romance are tightly intertwined, Indian captivity plays a starring role, and individual characters represent the various factions that would physically and symbolically compose the nation. And as in the other works I study in this book, gender and sex are matrices through which Mansilla can represent her indigenous protagonists and interpret the morality of their actions. While Mansilla can be not be claimed as a direct precursor to the scientific texts, *Lucía Miranda* highlights women's long involvement with the so-called Indian question and points toward the cultural and historical assumptions that would later ground more properly scientific projects.

As in the male anthropologists' later texts, gender is absolutely fundamental to Mansilla's racial vision in *Lucía Miranda*. In the novel, the Timbúes and their leader at first appear to be noble savages. They are "an inoffensive and tame people," naked and trusting (303). Marangoré is attractive, honest, skilled, and masculine without being threatening, safely married to a beautiful Indian woman (323). Meanwhile, the bad Indians in the text are characterized as hypermasculine and disrespecting of women, anticipating the vision of Zeballos and Mansilla. Siripo is ugly, reserved, and cruel, making him the archetypical barbarian.[27] When the Spanish-Timbú party finds an injured woman from another tribe, she pleads for her freedom so that she can care for her children. The barbarous Siripo begs to kill her while Sebastián, a representative of civilization, bandages her leg and protects her. Foreshadowing the novel's tragic denouement, savagery eventually overcomes civilization and Siripo hacks her to death "with a serene face and tranquil expression" (326). The juxtaposition between the brutality of the murder (and of a mother, no less) and his calm demeanor highlights

Siripo's savagery. Just as in the later accounts studied in Chapter One, in *Lucía Miranda*, treatment of the "inferior" sex becomes powerful proof of a man's civilization or savagery.

But in Mansilla's *Lucía Miranda*, Lucía is not just a victim. Rather, she plays an important role in the colonizing mission, intensifying and subtly shifting the liberal Argentine belief of the 1850s to 1870s that "civilization was a principle capable of feminizing the natural world, characterized by masculine brutality."[28] While the leaders of the expedition are Sebastián and other good, civilized men who create a homosocial bond with the Timbúes as they cross the pampas, Mansilla depicts Lucía as the true force of civilization.[29] In the end, what civilization the Timbúes receive from Sebastián and the Spanish soldiers is no match for their barbarous passion. Despite Sebastián's connections to Marangoré, the latter still breaks his promise and betrays the Spanish. Marangoré's ultimate betrayal demonstrates the limits of masculine civilizing attempts.

In contrast, Lucía's influence is effortless, profound, and enduring. She first proves women's stabilizing value by becoming a peacekeeper within the Spanish expedition, convincing the women to discourage their husbands from mutiny (300) and providing "solace and relief" to the suffering (306). Once in Argentina, Lucía is integral to the establishment of relations with the Timbú people. Although her husband and adoptive father are ostensibly in charge of the mission, they are highly dependent on Lucía since she learns the Timbú language with "prodigious ease" and acts as interpreter (307).

At the same time, Lucía forms relations with the Timbú women, taking advantage of their almost-automatic attraction to her beauty and grace to teach them Spanish ways (304). When Fray Pablo dies, Lucía's feminine spirituality and goodness make her the ideal replacement. She visits the *indias*' homes every day to instruct them in Christian doctrine and in how to be good women, insisting they "cover their nudity and try to observe in all their actions modesty and decency, which are women's most valuable charms" (316). In this way, Mansilla depicts the call for feminization of barbarous nature literally, tasking a woman with the role. The ease with which she develops these relations suggests the naturalness of women's participation on the frontier, a feminine sensibility that crosses racial lines and anticipates the trope of the civilizable woman that would later permeate Argentine anthropological writing.[30]

Mansilla's project also echoes another paradigm of the male liberals of the Generation of 1837, that of the republican mother. As in other works by Mansilla, in *Lucía Miranda* barbarism and conflict are depicted as the result of insufficient parental authority.[31] The Timbú mothers had previously encouraged their children to be rebellious and disrespectful, "judging that maternal love consisted of permitting the most rude and shocking acts" in order to prove the children's bravery. This practice horrifies Lucía, and she encourages the mothers to teach their children to be respectful and submissive from a young age (316). The nineteenth-century ideal of civic motherhood viewed the primary role of the mother as educating her children to be future citizens; in this passage, Lucía imposes this ideal on the Indian women as she encourages them to raise their children to be obedient. By educating Indian mothers, Lucía plays a fundamental role in shaping the future Argentine nation.

Her maternal role goes beyond teaching Timbú mothers how to properly parent. She actually functions as the surrogate mother for the wayward *hijos del desierto*, providing the civilizing education that the pampas could not. Indeed, as a text of the late-nineteenth century that reimagines the mythical founding of the nation, *Lucía Miranda* is rather odd in its focus on Spanish failure. The protagonist and her husband die before having a child, leaving no biological progeny to construct the future nation. The foundation appears in ruins, with nothing built on it. But although Lucía and Sebastián are childless, Lucía has not been un(re)productive: Mansilla represents the Timbú women as her surrogate children, molded by her to act as future citizens.

One of the clearest descriptions of Lucía's efforts in Argentina depicts her teaching Christian doctrine, surrounded by Indian women transfixed by her words. The *indias* are compared to "the child who repeats his first prayer, taught by his mother, and without realizing it feels, in the depths of his soul, mystic revelation that rises from the heart to the face, illuminated with a heavenly reflection (316). The Indian women, timid and innocent as the child, are taught by their mother Lucía to pray and love God. By connecting to them on this emotional, maternal level, she is able to change their hearts and minds, leading them closer to civilization. Reversing the image of the (male, celibate) Catholic missionaries teaching indigenous people in the wilderness, it is precisely Lucía's maternal potential that allows her to instill Christian virtues in the savages. In the face of these errant natural mothers, the republican mother of Argentine literature crosses racial lines, prepared more by race and spirit than actual parenting experience to set the children of the desert on a proper path.[32]

It is notable that many of the women in this passage are holding their own children, emphasizing a sort of matrilineal path of knowledge that will be repeated and foreshadowing Moreno's photographs of Tehuelche women clutching children while wearing rosaries. As Masiello notes,

"Lucía's project appears designed to inculcate a sense of republican motherhood in the Indian women she addresses."[33] The physical and epistemological reproduction inherent in motherhood suggests that Lucía's feminine civilizing mission will have long-term positive effects on the region.

Mansilla's vision for women is not particularly radical, as she doubles down on the importance of motherhood and the complementary relationship between the sexes. While essential to the civilizing efforts, Lucía is still intimately tied to Sebastián and there is no argument that she could go it alone. In this regard, Mansilla's philosophy is in line with that of other advocates of what María Rosa Lojo has called *feminismo discreto* (discrete feminism). Part reflection of the early steps of a movement, part political tactic to avoid outright objection, this vindication of women's rights focused almost exclusively on the domestic sphere.[34] Proponents emphasized the importance of women in society and argued the need to grant them more education and agency, without questioning the patriarchal structures that shaped creole society. Similarly, Mansilla gives Lucía agency and purpose in the building of Argentina but works within existing gender structures. Lucía's effectiveness on the frontier is through motherhood, not in spite of it. It may even be more effective than the masculine homosocial civilization that has no mechanism for reproducing itself.

Nonetheless, Lucía's presence on the frontier poses a paradox: while as a white woman she is essential to the civilizing project, as a sexual body she is a harbinger of disaster. Despite her positive effects on the indigenous women, it is Marangoré and Siripo's barbarous lust for Lucía that leads to the death of the Spanish colonizers. Lucía (and, by extension, white women) is thus portrayed as a necessary component of Spanish/Argentine plans in the River Plate and also a potential risk. In order to solve this problem, Mansilla continues to stress the role of women while limiting the influence of men.

The scene of Lucía teaching Christianity to the Timbú mothers, contrasted with the very little contact she has with the Timbú men, makes clear early on that *Lucía Miranda* is just as interested in civilizing women as it is in women as the providers of civilization. The relative value of the gendered Timbúes is further developed when the betrayal of Marangoré and Siripo is contrasted with the loyalty of Anté, a Timbú woman who is the first and most devoted of Lucía's Indian children. Anté frequently goes to the Spanish fort and allows the women to dress her up in European frippery, shedding her indigenous garb. Her transformation goes deeper as she also adopts Catholicism, quickly learning the Lord's Prayer (314). From outward appearance to inward devotion, little by little Lucía and the Spanish women shape Anté in their image. As a result of this transformation, Anté becomes Lucía's second-in-command and helps with her scheme to reveal the witch doctor's fraud. This scene makes clear that Anté is no longer one of the Indians, but something else entirely. As the tribe watches the witch doctor's machinations in fear, Anté looks on "with a tranquil face and serene expression," forming a striking contrast to the agitation of the other Timbúes (331). The physical contrast between Anté and the other women demonstrates how Lucía's civilizing actions have othered the Other from her own tribe.

This particular description also echoes almost exactly the language used to describe Siripo's face as he murdered the Charrúa mother. The use of the same words to describe Anté's confidence in rational Christianity and Siripo's attitude toward murder highlights feminine susceptibility to civilization and masculine distance from it. Even Marangoré, ostensibly the more civilized of the brothers, only partially accepts Spanish ways. Lucía invites him and his wife, Lirupé, to visit Spain, promising that they could all live together. Marangoré refuses, insisting that "the son of the pampa will never be able to breathe freely in the narrowness of your rooms" (337). The contrast between Anté and Marangoré/Siripo makes clear that as white women are the ideal civilizers, so too are women the susceptible sex when it comes to civilization. From Mansilla's perspective, it would take a more feminized society to solve Argentina's problems.[35]

Because indigenous women are extremely receptive to white women's civilizing efforts, they can be included in the nation. In the novel, Anté falls in love with a Spanish soldier named Alejo. He too is one of Lucía's adoptive prodigy, for she had promised his mother she would watch over him in Argentina (286). When Lucía discovers that Alejo and Anté love each other, she does not condemn the mixed-race relationship, but rather carefully prepares Anté for marriage, as a mother would her daughter. In this process, Anté's role as Lucía's progeny in the Americas is cemented, for "as time passed, the Spanish woman's heart transmitted a portion of its delicate perfume to the young Indian girl" (318). Anté carries forth Lucía's essence, like the biological child would a parent's eyes or disposition. Given that Alejo is also surrogate mothered by Lucía, he too carries her teachings, emphasizing Lucía's all-encompassing role in the founding of the nation. When Lucía and Sebastián are thrown on the pyre, Anté faints with emotion and is carried off by Alejo. As they escape the horrors of the Timbú camp, the third-person omniscient narrator reflects, "Where will they go? Where will their love find shelter? The whole Pampa offers them its immensity!" (359). Although Sebastián and Lucía have failed to found a purely European nation, Lucía has paved the way for Alejo and Anté to found a nation based on Christianity and Spanish values in the River Plate territory. Through the Alejo/Anté plot line, Mansilla's narrative not only reaffirms indigenous women's potential to be civilized, but also foregrounds racial mixing as a strategy for eradicating barbarism and thus establishing the base of the Argentine nation.

This acceptance of (certain types of) racial mixing is reinforced by the fact that Lucía herself is a mestiza, although of the Reconquista-era Peninsular variety. She is the daughter of "don Alonso de Miranda, young Spanish gentleman, poor and brave" and a Moorish girl (158–59). Contrary to many nineteenth-century intellectuals' beliefs that racial mixing would lead to racial anarchy and poor political and economic outcomes for the nations composed of hybrids, Mansilla's mestiza is an ideal woman. Like Homberg's Lin-Calél, Lucía is beautiful, intelligent, and Christian, a friend to all, and the most faithful wife imaginable.

Her personal history provides a key to Mansilla's vision of successful mestizaje. Lucía would rather die than submit to Siripo but is the product of a Spanish father and a racially other mother, just as any children of Alejo and Anté will be. In both Lucía's backstory and the novel's present, Mansilla insists that "inferior" woman can and should be incorporated through mestizaje, while racial mixing between an Indian man and a Spanish woman would be unthinkable. Men indelibly bear the stamp of barbarism, ready to betray their allies at the sight of white woman's body, but women in *Lucía Miranda* can be shaped into Christian citizens. Pairing them with Spanish males could further improve the nation, diluting the Indian's biological contributions just as Lucía strips away their cultural characteristics.

From Fiction to Reality: Mansilla's Critique of US and Argentine Racial Projects in *Recuerdos de viaje*

While Mansilla's feminizing and maternal vision of civilization meshed well with the dominant liberal ideologies of the 1860s, by the mid-1870s both text and policy reaffirmed masculine projects. As Adriana Novoa has argued, "Since according to the Darwinian theory, what was important in the struggle for existence was individual fitness, the traits necessary for survival were clearly masculine."[36] Rather than the gentler forces of republican motherhood and racial mixing, sentiment had turned to virile military action, emblematized in the Conquest of the Desert and the scientific-political tracts studied in Chapter One that generally depicted the struggle as that of properly masculine men against excessively masculine or excessively lazy barbarians.

Interestingly, Mansilla returned to many of the topics of *Lucía Miranda* in her travelogue, *Recuerdos del viaje*, published precisely in this period during the Conquest of the Desert. She had spent approximately four years in the United States, arriving for the first time just as the Civil War was beginning. With several children in tow, Mansilla visited New York, Washington, DC, Baltimore, and other locations along the East Coast. She published *Recuerdos de viaje* nearly two decades after her visit. Reflecting both Mansilla's additional experience and the development of Argentine scientific thinking, *Recuerdos de viaje* engages more expressly with racial science. She juxtaposes the Saxon and Latin races with regard to physical and psychological characteristics, noting the differing effects of food and hygiene on each (24, 86, among others). She also mentions Louis Agassiz and calls Santiago Arcos (her brother Lucio's interlocutor with regard to the unity of the human species) and Tullio Verdi (homeopath and Special Sanitary Commissioner in Washington, DC) her good friends (67, 122, 117).

In the text, she sharply criticizes the United States government for its Indian policy. The masculinity of colonial projects is not as immediately evident in *Recuerdos de viaje* as in *Lucía Miranda*, but Mansilla's targeting of government policies and institutions from which women were excluded once again links gender to exclusionary and inefficient nation-building practices. The Yankee, Mansilla claims, was not content to buy land from the natives but "decided to destroy the race using all the methods at his disposition. Death, betrayal, and robbery have been the arms with which they have combated them, promises and trickery their policies toward the children of the desert" (33). To highlight the hypocrisy of these policies, she quotes the *Times* of London's Spencerian declaration that there are two types of justice, one for the strong and another for the weak. Mansilla additionally claims that the Indian Department's only mission was "to get rich, robbing without shame, the pittance of the few Indians that still remain. . . . The Government knows it, tolerates it; I will say more, they approve it; and when they want to protect a good friend, they name him a delegate of the Indian Department" (33).

Although she does not name Argentina in *Recuerdos*, using instead "our country," "our race," "here," and "we," the implicit comparison is always present.[37] The Argentine Conquest of the Desert was ongoing at the time she was preparing and publishing her travelogue, and it is difficult to imagine that she did not make the connection between those events and her experiences in the United States. Furthermore, her use of the Argentine phrase "children of the desert" to refer to the Native Americans further links the native inhabitants of both countries.

Letters written by Miguel Malarín, the military attaché who studied US indigenous policy, to General Julio A. Roca in the first half of 1879 also suggest that Mansilla's commentary on the US reflected not just her own observations in the 1860s, but also knowledge of contemporary policy and plans to apply it in her own country. After his time in Washington, DC, Malarín traveled to Paris. The letters he wrote Roca from that city discuss plans for the Conquest of the Desert and the applicability of US policy to that campaign. Eduarda Mansilla and her husband, Manuel García, were in Paris at the same time, and Malarín writes of extensive conversations with García about indigenous affairs, including their anxious wait to receive a copy of Zeballos's newly published *La conquista de quince mil leguas*. García, Malarín indicates, was "a friend of Roca," supporting plans for the Conquest of the Desert.[38] Malarín also knew Mansilla well, and one of his final letters presents her to Roca as "a woman of exceptional character . . . educated and intelligent."[39] Given these circumstances, it is not unlikely that Malarín told Mansilla about his experiences in the US, either directly or through her husband, or that she could have participated in the triangular debates between Malarín, Roca, and García regarding the application of those policies to Argentina. It also seems highly probable that she would have read *La conquista de quince mil leguas*, either while still in Paris or once back in Buenos Aires.

Recuerdos de viaje was published less than three years later. Read in this context, it is clearly a rejoinder to those debates as well as the Conquest of the Desert carried out in the intervening years. Malarín admired the Bureau of Indian Affairs, and Estanislao Zeballos recommended that Roca and his troops "go ahead," fulfilling an Argentine sort of Manifest Destiny. In contrast, Eduarda Mansilla explicitly critiques this model for being cruel, corrupt, and ineffective. While her gender did not allow her to participate directly in making decisions about policy or military action, writing *Recuerdos de viaje* provided her with a space in which to continue the call for tolerance started in *Lucía Miranda* and to express her displeasure with the choices the men around her were making.

In *Recuerdos de viaje*, Mansilla also compares the US and Argentina in the context of racial mixing and its effects on a population. She claims that while the Saxons have mixed with black men and women, "they have maintained their distance from the Red Skins, with an antipathy worthy of concern for anthropologists, and that must undoubtedly have a serious physiological reason" (34). In line with her argument for (certain forms) of miscegenation in *Lucía Mansilla*, in *Recuerdos de viaje* Mansilla pathologizes the US tendency not to mix. She points out that many theorists attributed the United States' strength and success to this resistance to miscegenation but admits that she is not convinced. Instead, she casts this separation in a negative light, describing the indigenous chieftains as dignified, deserving of respect, and "deprived, disdained, deceived by men who profess a religion of equality and meekness, and who, however, do not practice the principle of their precepts: fraternity" (34). What some racial scientists called success, Mansilla sees as hypocritical and inhumane. Notably, unlike in *Lucía Miranda*, in *Recuerdos de viaje* Mansilla extends her support to the male chieftains. Whether this newfound sympathy for masculine indigenous peoples reflects a change of heart, the lesser tension of speaking of a foreign context, or a rhetorical move is unclear.

The implicit comparison with Argentina resurfaces here, for the United States' racial purity as a cause of its success was most frequently contrasted with the example of the racially mixed and chaotic Latin American republics. By questioning US policies, Mansilla defended her own country against racial maligning and the disdain of the North Americans who had co-opted the name "American" and who "believe only in themselves, and do not even give faith, nor do justice, to the real progress of our [South America's] Republics. We call them with a certain naivety, Northern brothers; and they ignore even our political and social existence" (38). She thus can use the example of the United States to both implicitly criticize the military destruction of the Conquest of the Desert and reaffirm the importance of the racial mixing that had long characterized Argentina's interior.

In *Recuerdos de viaje*, Mansilla continued the critique of masculine civilizing projects begun twenty years earlier in *Lucía Miranda*. Although not able to participate directly in policy making, she uses her writing to condemn masculine actions, intellectuals, and policies, including those closely tied to anthropology and the development of racial science. Unlike *Lucía Miranda*, *Recuerdos de viaje* does not lay out a specific contestatory feminine project. Nonetheless, she does continue to promote nation building on the basis of mestizaje, a process that necessarily required women's participation both in order to reproduce and in their role as mothers who could educate mestizo children to be good future citizens.

Mansilla's two works not only show that women had opinions about both colonization and racial science but also bring into focus one of the arguments of this book: that racial science and indigenous policy were never as unanimous, inevitable, or black and white as they might appear. *Lucía Miranda* anticipates the representation of hypersexual indigenous men and the association between women and civilizability seen in nearly all the anthropological texts of the 1870s and 1880s. This reinforces the idea that the concepts of civilization, barbarism, and racial change were based on cultural factors and economic/territorial concerns with roots that began far before the era of scientific racism. In the case of the River Plate, many of these factors were related to gender, particularly as prescribed by Roman Catholicism. As Mansilla's interpretation of the legend shows, it is the cultural association of women with Christianity that paved the way for their assimilation. Later scientific projects continued to rely on this linkage. Moreno's depictions of Tehuelche women in the museum showed them wearing crosses, and in Holmberg's *Lin-Calél* it is precisely Christian grace that sets Lin-Calél apart from the tribe.

At the same time, Mansilla's call for female-led, maternal projects of civilization, in agreement with gendered ideals prominent among the members of the Generation of 1837, clashes with the masculine visions of force of Zeballos, Roca, and the early works of Lista in the 1870s. While certain elements of her (pre-racial science) representations of race and gender persisted in their texts, others were replaced by new ideologies. At the same time, both *Lucía Miranda* and *Recuerdos de viaje* present interesting resonances with some post-Conquest of the Desert texts such as Lista's 1894 *Los indios tehuelches: Una raza que desaparece* and Holmberg's *Lin-Calél*. Like Mansilla, both men criticized military actions and argued for racial mixing as a way for improving Argentina, while never denying the Indian's fundamental lack of fitness for the modern world. While I am not arguing for direct influences between any of these texts, it is important to note how including an author like Eduarda Mansilla highlights the complexity of ideas in the period as different strands of racial theory, gendered ideals, religion, political context, and personal factors intertwine.

Feminine Colonialism and Florence Dixie's British Projects in Patagonia

Like Mansilla, Lady Florence Dixie had an upper-class childhood, international perspective, and genre-spanning professional writing career. Born in Scotland in 1855, Dixie was a seasoned traveler, a reporter, and a staunch defender of the equal abilities of women. She was a war correspondent during the Boer War in South Africa, and these experiences led to a popular travelogue, *In the Land of Misfortune* (1882). Many of her activities were specifically directed at promoting women's rights and public participation. She was a vocal advocate of women's suffrage and the president of the

British Ladies' Football Club.[40] These preoccupations are reflected in her writing, most explicitly in the feminist utopia *Gloriana, or, The Revolution of 1900* (1890), which argues that only true equality for men and women could cure society's ills and promote national progress.[41]

Two of Dixie's least studied works are children's novels based on her personal experience traveling in Patagonia: *The Two Castaways* and its sequel, *Aniwee, or, The Warrior Queen.* Like Mansilla in *Lucía Miranda,* in the novels Dixie explores the pampas and relationships between indigenous and white women. In contrast to Mansilla, however, she uses the material for elaborating a gendered and racialized vision of British agency in South America. In *The Two Castaways* (1890), fourteen-year-old British twins Margaret (Topsie) and Harry Vane are sailing to Santiago, Chile, to meet their father when they are shipwrecked off the Argentine coast near San Antonio. On shore, the two children are taken "prisoner-guests" of the Tehuelche Indians, who are led by Gilwinikush (99). Topsie and Harry learn the Tehuelches' ways and befriend Aniwee, the chieftain's daughter. Throughout, Topsie proves herself to be equal or superior to her brother in strength, intelligence, and character. She becomes Aniwee's friend and mentor, and with her encouragement Aniwee is named huntress and warrior of the tribe. Together with the Tehuelches and the Araucanians, led by the chieftain Cuastral and his son Piñone, the children defend the camp against Argentine incursions and survive a puma attack, a prairie fire, and severe dehydration. On an expedition deep into the forest, they meet an exceptionally old hermit. To the surprise of all, he is the twins' great-great-granduncle Sir Harry Vane, who was shipwrecked many years ago and had forged a life in Patagonia (269). He married the daughter of a British missionary, had a child, and discovered an impossibly rich gold mine, the Mine d'Or, where all but he perished at the hands of Andes demons (297). After meeting Sir Harry, Topsie and Harry return to the Tehuelche camp where their father and uncle find them and prepare to take them back to England with the Tehuelches' blessing. The novel ends with young Aniwee's marriage to Piñone, the brave Araucanian warrior.

Also published in 1890, *Aniwee, or, The Warrior Queen* is the sequel to *The Two Castaways.* In the two years between the novels' action, Aniwee gave birth to a girl, and Piñone and Cuastral were believed to have been murdered by treacherous Argentines. Because Aniwee has impressed the Araucanians with her "sagacity, courage, and devotion to their interests," they named her baby daughter, Guardia, the next cacique, with Aniwee as Queen-Regent (2–3). Topsie and Harry return to Patagonia to visit.

The story follows two intertwined plotlines set against the backdrop of Argentine-Indian conflict. In the first, a jealous Araucanian named Inacayal plots to kidnap Aniwee's daughter, thus gaining power for himself.[42]

The novel pivots to the second narrative arc when Topsie is thrown from her horse and taken captive by a hairy man who looks like an ape. A Dixie-invented combination of numerous myths, this "Trauco" lives in a haunted forest rumored to be the site of the legendary Enchanted City of the Caesars (53, 74). Among the Traucos, Topsie discovers that Piñone and Cuastral are not dead but rather fellow prisoners, similarly victims of Inacayal's treachery. The stories merge when Aniwee and the Vane family rescue Topsie, Piñone, and Cuastral at nearly the same moment that Guardia is kidnapped and abandoned in the forest. While searching for the baby, Harry, Topsie, Aniwee, and the rest of the rescue party return to the Mine d'Or visited in *The Two Castaways*. There, an earthquake buries them alive. They survive only because of Topsie's ingenuity and bravery in climbing the sheer rock wall (232). After this daring escape, the party returns to the Tehuelche camp, peace is celebrated with the Argentines, and all live happily ever after. As I will discuss later, both novels rely on the benign colonizing powers of the British to arrive at their happy endings.

As the British worked to cement their imperial projects in the late nineteenth century, the market for juvenile fiction exploded.[43] One particularly popular strand of literature was the heroic adventure, in which children and adolescents undertook grand excursions in novels that paralleled imperial processes.[44] Set in prospective colonial spaces around the world, these texts were often racist. Authors such as H. Rider Haggard, Rudyard Kipling, and GA Henty depicted native inhabitants as barbarous and needing the British to save them from their own inferiority. As the title of Bristow's work *Empire Boys: Adventures in a Man's World* reveals, these novels featured young men and were primarily marketed toward them. Perusing the catalogs at the back of the 1890 John F. Shaw edition of *The Two Castaways*, the section "Shaw's Our Boys' Bookshelf" is populated with titles such as *Emperor's Boys*, *A Real Hero: A Story of the Conquest of Mexico*, and *Winning an Empire*. In this kind of title, race, nation, empire, and manliness are clustered together within the ideology of imperialism.[45] In comparison, the booklists oriented toward girls feature titles such as *Two Little Boys, or, I'd Like to Please Him* and *Lotty's Visit to Grandmama*, focusing largely on the domestic, both home and homeland. Although these texts linked domestic spaces to imperialism through the importance of preserving the integrity of the race, adventure novels featuring girls as protagonists were rare.[46]

Dixie's novels also intervene in the masculine world of exploration and science. Although she does not explicitly acknowledge her debt to Musters's *At Home with the Patagonians*, references to the English explorer's account of his year living among the Tehuelches appear throughout *The Two Castaways* and *Aniwee*. At the most basic level, Dixie borrows people and place names: Musters's Cuastro becomes Cuastral, "Aniwee" appears in his Tsoneca-English glossary, and the real-life Inacayal (described by Musters and later photographed by Moreno) lends his name to the jealous villain in the second novel. Musters's discussion of the mythical Traucos and Ciudad Encantada informs both novels and becomes a major plot point in *Aniwee*.[47] And as I will show later in this chapter, Dixie reproduces certain scenes and dialogues from Musters's work almost exactly. *The Two Castaways* and *Aniwee* are thus highly unique works, representing not only amendments to the British juvenile adventure novel but also feminist rewritings of one of the most canonical texts of Patagonian exploration.

In the novels, Dixie elaborates a world where British girls shared the adventure and the imperialist responsibility with their brothers. In order to create this vision, Dixie turned to the best evidence she had for women's equality: herself. Topsie, the fourteen-year-old heroine, is Dixie's fictional double, collapsing several periods of her life into one dynamic character. Like Dixie, Topsie has a twin brother. She is energetic, strong, and talented at hunting, riding, and wilderness skills. She is a better shot than her brother (*Two Castaways* 75), participates equally with the men in hunting and setting up camp (58), and is often fearless while her brother has doubts (26). In her autobiographical writing, Dixie similarly depicts herself outlasting the men in her company. For example, in *Across Patagonia*, she describes herself as a hardier horsewoman than the sailors of a German ship who accompany them on a day's excursion. Returning to town after many hours of riding, Dixie notes, "they were most of them completely done up," whereas she immediately mounts her horse to return to camp without complaint, then rises early the next morning to pack up and begin again (*Across Patagonia* 54).

Reflecting Dixie's own upbringing, Topsie's skills are attributed to the fact that she had been given an education similar to that of her brother, and Harry supports Topsie's activities just as Dixie's brothers did hers. This leads to one of Dixie's frequent moments of moralizing, as the narrator of *The Two Castaways* reflects on Topsie's abilities and recommends:

> Those boys who are apt to despise their sisters as "only girls," and regard them in consequence as weak and feeble, should bear in mind that they are so because of their training and education; and in following the adventures of Topsie and Harry, take to heart the simple fact that a girl can be every inch as plucky and enduring as a boy, if she gets fair play in her early training. Generous-hearted boys should bear this in mind, and do their best to help their sisters get fair play, even as Topsie did. (52–53)

Topsie also gives voice to Dixie's certainty that the Patagonian territory was a freer space for women than Great Britain. Argentine men such as Lista and Mansilla wrote extensively of their desire for the freedom of the other side of the frontier and the ease with which one could slip away from the trappings of civilization. This liberty was intensified for elite women who looked to break free of the bounds of both civilization and gender. Dixie's fellow female British travelers—including Mary Wollstonecraft, Harriet Martineau, Isabella Bird Bishop, and Mary Kingsley—all wrote of the desire to escape people and explore solitudes, excusing their unusual actions by exaggerating their distance from England in their writing.[48] In *Across Patagonia*, Dixie hints at this appeal, opening her account with a celebration of the emptiness of Patagonia: "But nowhere else are you so completely alone. Nowhere else is there an area of 100,000 square miles which you may gallop over, and where, whilst enjoying a healthy, bracing climate, you are safe from the persecutions of fevers, friends, savage tribes, obnoxious animals, telegrams, letters, and every other nuisance you are elsewhere liable to be exposed to" (2-3).[49] The very fact that the imperial frontier was difficult to access paradoxically created spaces where those who did arrive had the latitude to explore new ways of being.[50]

Patagonia's relative freedom, even for a woman of unconventional upbringing, is driven home in the *Two Castaways* by a dialogue between Topsie and her father. Admiral Vane asks Topsie how she will feel about wearing petticoats again when she returns to Britain. She replies, "I shall hate them, father. . . . I've been pretty free all my life in comparison with other girls, but I never knew what it was to feel really free till we were wrecked on the shores of Patagonia" (373). Despite its challenges, Patagonian space is depicted in the novel as equal or superior to civilized Britain, precisely because of the greater freedoms that the barren landscape offers for women. On the frontier, far from the hegemonic center, Dixie and her fictional double find the freedom to fully express the potential developed by their unconventional educations, innate strength, and intelligence. The return of Topsie's father, the patriarch, marks a return to the still-limited spaces of enlightened, civilized Britain.

In both novels, Dixie replicates the equality between Topsie and Harry in Tehuelche society, and this decision leads to more positive representations of the Tehuelches in general. Like Topsie, Aniwee has a somewhat open-minded father and proves herself to be equal to men, earning the right to rule her tribe. The two novels together trace Aniwee's slow recognition of the inequalities that surround her and her first steps toward making change. After her successful participation in the hunt, Aniwee sees clearly the wrong in which they have all been living:

> Poor little Aniwee! In that child's mind a great revolution against wrong was uprising. She had seen the women toiling, while the men feasted and hunted and fought, and enjoyed all the excitements and pleasures of Indian life. In her small wee heart she resented this great inequality, and, like all pioneers, determined to do and dare all to win freedom for herself and her sex. (*Two Castaways* 133)

In this passage, the narrator highlights Aniwee's youth and inferior position (child's mind, wee heart), and contrasts them with the great things of which she dreams. Being small or feminine, the narrator insists, does not equal weakness or inferiority, and the narrative aims to inspire its young readers to equally "do and dare all to win freedom."

In her position as Queen-Regent, Aniwee takes on the role of social reformer and argues for the equal participation of women. A "pioneer," she allows other girls to participate in war, inspiring them to follow her into dangerous situations when the men were too scared to proceed (*Aniwee* 139). She rebukes her father, Gilwinikush, for claiming to be childless due to the death of his only son. She reminds him that he still has her, and, though she is a girl, "a girl can be as brave, aye, braver, than a boy; a woman as great, aye, greater than a man." She then repeats the criticism seen in the Argentine scientific texts with regard to the gendered distribution of labor in Tehuelche life. Dressed in warrior uniform, the ironically labeled "mere girl" states,

> My brothers applaud the deed of Aniwee, yet wonder at it because she is a woman. But why should they so wonder? Has not a woman the right to do great deeds? Must she for ever toil and do the work of slaves, while the man plays that part in life which is pleasant and agreeable? Aniwee thinks this should not be so. She thinks the burdens of the toldo should be shared alike by man and woman, and that the woman should taste some of the joys of life. Therefore has Aniwee acted as she has done, and will continue to do, to prove that a woman can be both a warrior, a hunter, or cacique, and yet be a woman still. (*Two Castaways* 217–18)

The male warriors almost immediately accept her once she states her case, and Aniwee quickly becomes "the idol of the Indians" for her "wisdom, foresight, and pluck" (*Two Castaways* 168), living proof that women are men's equals. This quick acceptance of her rapidly changing position in the tribe reaffirms Dixie's belief that such equality is only natural. Its logic is so clear that even uncivilized peoples accept it without complaint.

In Dixie's travelogue, *Across Patagonia*, she meets a "real Patagonian Indian" whom she uses as inspiration for several ethnographic-style paragraphs. Although Dixie does not reference any sources, her portrait of the Tehuelches is very similar to those drawn by Moreno, Lista, and Zeballos. She writes, "they are extremely lazy, and, plentiful as game is around them, often pass two or three days without food rather than incur the very slight exertion attendant on a day's hunting. But it is only the men who are cursed or blessed with this indolent spirit. The women are indefatigably industrious. All the work of Tehuelche existence is done by them except hunting."[51] This passage evokes numerous segments of the anthropologists' work, but perhaps most explicitly Lista's assertion, "The usual occupation of the men is hunting, although as I have already said, they are so indolent that they tend to pass entire days without having food to eat. The women are, in contrast, active and industrious."[52] Given that Dixie, Lista, Moreno, and Zeballos had all read Musters's *At Home with the Patagonians*, it is likely that some similarities between their works reflect this shared knowledge base. Furthermore, it is not impossible that Dixie would have heard of Lista's work, as *Viaje al país de los tehuelches* was published in March 1879 while she was in Argentina. She also would have encountered similar ideas in newspaper articles, conversation, and English-language texts about Argentina; *The Times* of London even reported updates on General Roca's campaigns in Patagonia.[53]

Dixie's children's novels further develop the oppositional characterization of Tehuelche men and women, again using the language of anthropology. Aniwee gives voice to the observations Dixie herself had made on her journey across Patagonia while her disdain for the near slavery of the women and the limited responsibilities of men recalls the words of Moreno, Lista, Lubbock, and others. In contrast with the male anthropologists' texts, however, Dixie does not use this characterization to condemn the Tehuelches as barbarous or uncivilizable, but rather depicts these gender relations as part of a universal system of oppressive patriarchy affecting European women as well as indigenous women in Patagonia. Although including ethnographic elements, the fictional space of the children's novel gave Dixie more latitude to develop a sympathetic portrayal of indigenous people than the distance a true ethnographic text might, a dichotomy that Fiona Mackintosh observes when comparing Dixie's novels to her own more scientific travel writing.[54] Dixie's two novels are also a sharp departure from the racist ideology of most boys' adventure novels of her time. As I will demonstrate shortly, these positive representations of indigenous peoples stem at least partially from Dixie's efforts to prove women's worth.

The language of "sisters" and "friends" peppers the novels, reminding the reader that despite different origins the two girls share many experiences and struggles. In one particularly clear passage, Topsie and Aniwee are described as racing across the pampas, side by side. In this "striking scene . . . there was the white girl in the boy disguise, giving the lie direct to the statement that the woman is the inferior of the man; and there beside her rode her coloured sister, eager and willing to bring the same false statement to book" (*Two Castaways* 128). One dark, one light, but two women united by age, ability, and ambition, unwilling to continue following social convention.

A conversation between them allows Dixie to explicitly critique British society. In the second novel, Topsie explains to Aniwee that in "the great white land"—Britain—women are not slaves as among the Tehuelches, but there were still limits on their behavior, such as their exclusion from Parliament and the military (*Aniwee* 15–16). She speaks of the need to change laws and highlights Aniwee's becoming Queen-Regent as a positive example of the maxim that "great changes often come quickly" (*Aniwee* 16).

As the above example indicates, the literary and ideological demands of Dixie's feminist project create space for her indigenous characters to be represented quite positively as full humans, brothers and sisters able to progress and adopt modern ideals. In this regard, she draws a clear parallel between American Indians and women. Whereas masculine British colonial texts linked indigenous people and women through their weakness and inferiority, Dixie highlights their shared potential.[55] She explains their historical marginalization and perceived inferiority not as the result of biological shortcomings, but rather as the consequence of a lack of proper education. Just as Dixie's narrator expresses early in the first novel that "a girl can be every inch as plucky and enduring as a boy, if she gets fair play in her early training" (*Two Castaways* 53), by the end of the second novel it is clear that the Tehuelches, too, possess this ability.

This characterization of the Tehuelches as willing and able to change leads to Dixie's second major project in the novels: a justification of British cultural influence in the Patagonian region. Although British attempts to gain formal control of the region ended in 1807, diplomacy and commerce irrevocably tied the two nations together throughout the nineteenth century. The 1825 Anglo-Argentine Treaty of Friendship, Commerce, and Navigation recognized the sovereign powers of the two nations while ensuring a free-market relationship that would encourage commerce while protecting the rights of expats in Argentina and the United Kingdom.[56] In the intervening decades, British investment in Argentine banks, railroads, and industries grew at a rapid clip, reaching 150 million pounds by 1890.[57] This amount made Argentina one of the most significant sites for British investment outside of the United Kingdom, on par with true colonies like Australia and Canada. Likewise, British products comprised over 50 percent of Argentina's imported goods in the late nineteenth century, making Argentina more dependent on the United Kingdom than any other country.[58] Studying this informal imperialism in Mexico and Central America, Robert Aguirre stresses the importance of cultural activities to maintain and strengthen commercial ties.[59] Museum displays, soccer games, and adventure novels provided the "ideological support" necessary for British commercial activities.[60] Published just months before the Baring Crisis significantly damaged Anglo-Argentine commerce, Dixie's children's novels can be seen within the framework of actions that aimed to shore up Britain's cultural and economic role in the region.

In her accounts of her travels in Africa, Dixie presents herself as "a potential agent of British imperial authority."[61] Her childrens' novels about Argentina further develop this project. An episode at the beginning of *The Two Castaways* locates the novels in an imperial context: in Rio de Janeiro, Topsie and Harry scale the Pao de Azucar and plant the Union Jack on top (*Two Castaways* 28). Their daring, intelligence, and skill equip them to do something Brazilians had deemed impossible. Physically above the city, they model British superiority. Topsie leaves a note in a bottle marking their claim: "It is here recorded that Margaret and Harry Vane have planted the British flag on untrodden ground, and annexed it for her Majesty of Great Britain and Ireland. God save the Queen" (*Two Castaways* 29). As Topsie and Harry steam out of Rio toward Patagonia, the passengers give three cheers for the British flag on the Pao, "that great emblem of the great British Empire" (*Two Castaways* 33). This episode, although presented as a frivolous child's joke, primes the reader to take seriously Dixie's claims for British influence in Patagonia, for she describes Patagonia as similarly untrodden territory and thus also up for grabs.

Once the children are in Patagonia with Aniwee and her tribe, Dixie's imperial argument becomes clear: the Tehuelches, like girls, need a proper education in order to fulfill their potential, and Topsie is just the woman to give it to them. Although she and Aniwee are depicted as sisters, sharing the bonds of womanhood, Topsie represents the voice of modern rationality that opposes Tehuelche primitivism. She benevolently and gently removes the Tehuelches from the errors in which they live, teaching Aniwee about language and science and lifting "the mists of superstition" from her eyes with regard to natural phenomena like avalanches (*Two Castaways* 246). In having Topsie explain these things to Aniwee, Dixie also justifies her own participation in the scientific enterprise. She too was able to understand nature and used her powers of observation and first-hand experience to contradict Darwin.

Additionally, Aniwee does not think to question her position in Tehuelche society until Topsie makes the suggestion. After Aniwee catches an ostrich, proving her hunting prowess, the translator Gonzales casually remarks that "by the Indian laws of hunting, if Aniwee were a man it would be hers" (*Two Castaways* 113). It is Topsie, not Aniwee, who defends the Tehuelche girl's right to the bird. In response, Gonzales brusquely insists that Tehuelche women do not hunt, but concedes to Topsie's demand anyway (114). In this scene, Topsie is the catalyst that transforms Aniwee's innate potential into power and provokes the monumental shift in Tehuelche culture developed throughout the two novels. Without Topsie and her education, Aniwee may never have thought to question the status quo.

In Dixie's novels, indigenous women share the potential of British women but need a civilized push from outside in order to capitalize on it. This conceptualization of the British girl/woman's role in Patagonia echoes nineteenth-century British feminists' understanding of white women's roles in civilization and development. "The white woman's burden," as Antoinette Burton has labeled it, called for British woman to fully embody their role as nurturers and educators in order to rescue the wretched women of the empire.[62] As in the case of Dixie's novels, in this fashion, women could play a fundamental role in the imperial project even while using the female Other to assess their own progress in the fight for women's rights.

Queen Victoria is the model that Topsie and Aniwee aspire to, representing both women's greatness and the ideal relationship to "inferior" peoples. A conversation about Victoria is one of the clearest signs of Dixie's debt to George Chaworth Musters: in Musters's account, an Indian named Tchang asks him about the "cacique of the English." When Musters explains that she is a woman, Queen Victoria, Tchang asks if she is married, if she is a mother, and if she is wealthy. He then rides on, exclaiming "A woman cacique! A woman cacique!"[63] In *The Two Castaways*, Topsie also explains to a surprised Cuastral that Victoria is the "head cacique" and "over the world her power is known . . . her warriors are fearless and brave." Whereas in Musters the conversation merits a brief mention as a surprising anecdote, Dixie develops the conversation, presenting the Queen as an ideal example for Aniwee. Topsie goes on to explain, "Indians and all manners and kinds of the red man and darker races own her beneficent sway" and that the

Queen wins their friendship by protecting them (*Two Castaways* 164). Thus, Topsie (Dixie's fictional double) is Queen Victoria's metonymic representative in Patagonia, tasked with educating and thus winning the favor of her darker sister, Aniwee. Together in Patagonia, under the influence of the British Queen and in adherence to the value of equal education for all, women and Indians can progress. Colonialism, scientific interactions with indigenous peoples, and women's rights cannot be separated, as the equality of women's suffering depicts the Tehuelches as ready for scientifically informed colonization, and the colonizing powers of Topsie (and Victoria) are a sign of their gender's equality.

Curiously, there had already been a real-life "Queen Victoria" on the pampas when Dixie wrote her novel. Musters described a visit from a group of Tehuelches including "an old squaw rejoicing in the name of 'La Reina Victoria' (Queen Victoria), who was the occasion of much chaff, my Chilian [sic] friends declaring I ought to salute the sovereign of the Pampas in due form."[64] Eight years later in *Viaje a la Patagonia austral*, Moreno wrote of a place called "*Pozo de la Reina* [Queen's Well], whose name comes from the fact that the Tehuelche Indian called 'Queen Victoria' by her compatriots had once fallen in it."[65] Later the same year, Lista noted that his group camped at *Pozo de la Reina* and described the origin of its name in nearly identical language to Moreno.[66] These two anecdotes raise more questions than answers: how had the Tehuelches learned of Queen Victoria? From a traveler or colonist pre-dating Musters? Did they read Argentine papers like Mansilla noted of Ranquel chieftain Mariano Rosas? What sort of power did the Tehuelche "Queen Victoria" have and how did her position support or clash with the explorers' assessments of gender roles?

In contrast to both Mansilla's text and traditional British imperialist works that portrayed women as responsible for the biological and spiritual protection of the race,[67] Dixie locates women's influence over colonized peoples as largely outside of motherhood and the domestic. The novels take place almost entirely outdoors, and Topsie and Aniwee participate equally in hunting and fighting. There are also notably few mothers in the novels. Both Topsie and Aniwee are raised by their fathers after their mothers' deaths. Topsie's teaching of Aniwee does not take on the maternal tones of *Lucía Miranda*, for they are the same age and engage as peers. The only real mother is Aniwee herself, and motherhood is just one of the roles she plays, especially given that she is separated from her daughter, Guardia, for most of the novel. Furthermore, Queen Victoria is not depicted as a motherly figure. Instead, Topsie and Harry's befriending of the Tehuelches is extrapolated to the imperial level as Topsie speaks of the Queen earning the friendship of her subjects. Dixie's books are thus radically different from both *Lucía Miranda* and other British novels of her time, both in the avoidance of motherhood and the tolerance of a girl engaging in so-called masculine behaviors.

The final piece of Dixie's push for British cultural influence in Patagonia is to juxtapose this benevolent, friendly approach with the perceived barbarity of the Argentine military venture. Given how quickly Aniwee and the Tehuelches "progress" under Topsie's guidance, Dixie implicitly attributes the Tehuelches' barbarism to Argentine neglect or mishandling. There are no Argentine women in the novels, symbolizing the lack of forces for civilization and love. The only representatives of Argentina are masculine warriors, who are cruel and ineffective in their attempts to control the Tehuelches. Gilwinikush complains of Argentine mistreatment (*Two Castaways* 95), and it is Argentine treachery that puts an end to Aniwee's wedded bliss (*Aniwee* 2).

In other moments, the Argentines are dehumanized, turned into animals like snakes (*Aniwee* 26) or foxes "intent on a brutal and cruel act" (*Aniwee* 22). These images echo Zeballos and Mansilla's descriptions of male Indians as wolves or other animals stalking their prey. Although the language is similar, Dixie shifts the group being modified, constructing the civilizer as a barbarian in order to introduce an alternative civilizing element, Britain.

In an even more explicit passage, after numerous Argentines are slaughtered the narrator observes, "Ghastly as the fate of their companions had been, it was but fit retribution for the many deeds of cruelty and murder that had been committed on the Indians" (*Two Castaways* 159). Interestingly, these are some of the same words that Eduarda Mansilla used to critique US Indian policy in *Recuerdos de viaje*. Once again, Dixie's words ring ironic for the reader familiar with Argentine texts, for these were also the very justifications used by some River Plate elites to destroy indigenous cultures. For example, Zeballos wrote in *Descripción amena* that the chieftain Namuncurá and his Indians "lived by theft and waged war against the Christian with implacable cruelty and hate. . . . Their invasions in our lands leave footprints stained with blood and marked by fire and looting."[68] Therefore, he writes, "It was necessary to fight the País de los Araucanos [Araucanian Region], toldería by toldería, stabbing their untamed warriors, taking their families captive, and evicting the nomadic populations from the open-country and forest lagoons, centers of life in the desert."[69] Dixie would certainly have heard similar rhetoric during her 1878 travels in Patagonia or in later publications, contemporaneous with General Roca's first offensive.

In her children's novels, Dixie appropriates this discourse to discredit Argentina and justify the imposition of her own colonizing culture. The horror of masculine colonization is contrasted with British colonialism, which Dixie characterizes as feminine and thus benevolent and enlightened. Over and over, Topsie and Harry insist they have nothing in common with the hated Cristianos and "their sympathy was entirely with the Indians" (*Two Castaways* 365). Topsie wonders aloud why the Argentines would not "leave these splendid fellows alone? First they steal their land, and then their wives and children; and when, as in common nature bound, the Indians rescue them, they are pursued like this. Why it is worse than in North America!" (157). The juxtaposition between Argentine cruelty and British benevolent guidance is further reinforced by the fact that Topsie and Harry are often referred to as the Indians' "white guests" or "white friends." Thus, the Argentine is depicted as an unjust, unchristian, bloodthirsty aggressor, whereas the "white friends" from overseas with the woman cacique are friendly and respectful. In this way, Dixie's children's books use the Patagonian frontier to simultaneously advocate for equal respect for women and British influence in the region.

Women, Racial Theory, and Colonial Projects

Mansilla and Dixie's texts clearly demonstrate one of the arguments of this book: although scientists were almost exclusively men in the period between 1860 and 1910, women were aware of the development and the deployment of racial theory and appropriated it for their own purpose. Racial theorists used indigenous women's bodies, captive women's bodies, and the imagined ideal creole woman to locate differences between races and rank them hierarchically. Additionally, concepts such as savagery, civilization, hybridity, evolution, and even nation did not exist without women's bodies. Their wombs were the site of reproduction and racial mixing, while the ways they were treated by men and the familial structures they participated in marked distinctions between groups.

Women also challenged current events developed in conjunction with racial theory. Looking back from the present, scientific racism often appears to have been *the* ideology of the late nineteenth century, shared by all and devastatingly effective. Nonetheless, some men and women—indigenous, mestizo, and creole—did challenge it, and racial science was far from unified in its beliefs and aims. Indigenous women such as Carmen and Clorinda Coile served as informants to Argentine anthropologists, shaping the representations of indigenous peoples in their scientific and nonscientific texts. Lucio Mansilla admits that Carmen is attempting to use sex and desire to manipulate him; it is unlikely she was the only one to attempt such an approach. The Tehuelche women taken to the Museo de La Plata also complicated men's scientific plans, giving different answers to the same questions and selling their weavings in town instead of allowing them to be displayed in the museum. While these are small moments of victory in a story that ultimately ended tragically for them and most Argentine indigenous groups, it is historically and politically important to recognize these disruptions. Finally, in their texts, Dixie and Eduarda Mansilla criticized men's scientific-political projects while also advocating for colonial strategies and national identities that were somewhat more inclusive of both indigenous peoples and women.

Just as there were differences among the racial theories and projects elaborated by men like Holmberg, Lista, Moreno, Lucio Mansilla, and Zeballos, there was also variation in how women resisted. In their position as white, upper-class women, Eduarda Mansilla and Dixie had power the indigenous women did not and were able to use writing to push back against civilizing projects. Race was not the only inflection point, however, for indigenous women had highly different interactions with science according to their social position within the tribe. In the Boote photographs from La Plata, women more closely related to the chieftains were less exposed, and we have more-complete (although still very fragmentary) records of their lives than those of the unnamed women from the lower strata of the community. Further differences in the photos and between Dixie and Mansilla's texts also show the importance of nationality, personal connections, and religion to the ways in which women used and were used by science. Sex and gender were tightly intertwined with racial science, and these gendered elements were always intersectional. Approaches to studying nineteenth-century race must be so as well.

CONCLUSION

An Enduring Legacy

The Nineteenth Century in the Twentieth and Twenty-First

The fate of the remains of Tehuelche cacique Inacayal's wife is perhaps the most pointed and emotional shorthand for this book's thesis. Brought to the Museo de La Plata with her tribe and photographed by Boote around 1886, she likely died sometime in 1887 while still living in the museum. Unlike others from her group, her death was not registered in Argentine papers and to this day she is nameless, remembered through only the sketchiest of details. Nonetheless, her body was studied by generations of anthropologists and was displayed for years for Argentine citizens behind a pane of glass physically marking her as separate from both the present day and the nation. Although hardly more than a shadow in the historical record, she was fundamental to the creation of the scientific and popular stories that cemented Argentina's reputation as a white, European nation.

As this book has argued, the development of this identity through science cannot be separated from women's bodies and their diverse roles as anthropological objects, ethnographic informants, symbolic spaces, wives, mothers, and sexual partners. Although the "proper" anthropologists were all male, white women were involved with scientific-racial projects too. Some were complicit with oppressive policies, such as those who taught indigenous women to be domestic servants or those in the Sociedad de

Beneficencia that organized and ran the distribution auctions of conquered women and children. Others, such as Eduarda Mansilla and Florence Dixie, resisted such projects or reshaped them to meet their own needs.

Indigenous women have been doubly forgotten from historical narratives due to their race and gender. Nevertheless, this book has shown that they were intimately connected to scientific projects. Like Inacayal's wife, their bodies were scrutinized, measured, and displayed. Others flirted with, slept with, or provided information to the scientists. Throughout, their interactions with the Argentine scientific world were marked by both racial and gendered paradigms.

Many Argentine writers argued that women were easier to civilize than indigenous men. At first glance this assertion is surprising, given the strong ties between citizenship, civilization, and masculinity in the period. Nonetheless, there were several reasons to presume women were more apt to be civilized. As seen in Chapter One, they were considered hard workers and thus more likely to contribute productively to society. The Hispanic concept of *marianismo*, which associated femininity with the spiritual, also meant their gender could predispose them to the adoption of Christianity. Indeed, we see this in Boote's pictures of the Tehuelche women wearing rosaries, the grace associated with Holmberg's mestiza protagonist Lin-Calél, and the ease with which Eduarda Mansilla's Lucía is able to step into the priest's role and win numerous female Timbú converts. Their role as mothers could also transform them into a link in a civilizing chain, as they could teach their children proper behavior.

Perhaps most importantly, civilized indigenous women presented no real threat to the power and control of white Argentine men. Because they were doubly subordinated by gender and race, even if they were culturally assimilated they would still be excluded from power due to their gender. Their place was in the home, their focus the domestic. Indigenous men, however, were legally citizens due to their birth in the nation, and their gender enabled them to flow fluidly through public and private spaces. Should they be civilized, they might demand to exert the rights of civilized men and thus challenge the status quo.

Although women were generally represented as less threatening than indigenous men, the scientists' own texts reveal moments of female actions at odds with their efforts to control them through observation, measurement, and even sexual activities. A counter-reading of Lucio Mansilla's text shows Carmen manipulating Mansilla even as he aims to manipulate her. In other texts such as those of Lista and Moreno, indigenous women served as gatekeepers, informants, and guides, potentially shaping their own representation in scientific materials. The photographic and textual archives of the Museo de La Plata reveal numerous moments when indigenous women acted or may have acted on behalf of themselves and their families. Even the idea of indigenous women as a force for civilization through generational teaching hints at the danger of this same power should they chose to perpetuate their own culture instead of that being imposed by the nation.

Of course, the fact that women and their bodies were used to disrupt interracial encounters does not necessarily mean that women held power within their own communities; indeed, the nineteenth-century texts are ambiguous regarding the extent of indigenous women's independence of action. Together, the presence of these women, both indigenous and white, challenges the idea that Argentine anthropology, colonialism, and identity processes were exclusively masculine spaces.

Furthermore, although Argentina has frequently been excluded from studies of the relationships between sex, racial theory, and colonialism due to its European reputation, it is not an exceptional case. Rather, it is precisely the relationships between sex and racial theory that gave rise to the colonial projects that cemented that white identity. For Mansilla and Zeballos, the hypersexual nature of the male Indian was one of the primary indicators of his savage status. Fears over hybridity, the physical reproduction of Indians, and the interruption of the national family provoked a desire to curb this sexuality, both through the outright violence of the Conquest of the Desert and through redistribution policies that isolated indigenous men from sexual partners.

While scientific texts could shape policy, in order to reach the broader audiences needed to affect the imagined community of the nation, they had to turn to other genres. Fictional texts, including Zeballos's *Painé* and *Relmú*, Zorrilla de San Martín's *Tabaré*, and Eduarda Mansilla's *Lucía Miranda*, took advantage of the fear of mixed-race unions to create suspense. Dramatized in prose and verse, the effects of the hypersexual Indian's actions on the protagonists intensified the anxieties of racial mixing and justified his exclusion from the national imaginary.

But just as sex caused repulsion, it was also a mechanism for bringing people and cultures together. In *Una excursión a los indios ranqueles*, Lucio Mansilla used the similarities of women and sexual practices everywhere to argue for the unity of the human species and fair treatment for indigenous peoples. Lista's attraction to the freedom of the frontier turned into a romantic relationship with a Tehuelche woman and a mixed-race child. This sexual encounter marked a notable shift in his scientific thinking, particularly with regard to the mechanisms of racial change and the incorporation of Patagonian Indians into the nation. Despite these approximations, however, racial mixture was never unproblematic. Lucio Mansilla's flirtations remained largely textual, and his greater sympathy for the Ranqueles dissipated in ensuing years. Lista's affair provoked a scandal in Buenos Aires that was at least partially responsible for his removal from his government position. Furthermore, as Lista himself noted in his *Los indios tehuelches: Una raza que desaparece*, his most sympathetic text, the Tehuelches were already disappearing. Even as Lista pleaded to physically protect them, he advocated for policies that would hasten their cultural assimilation.

Lucio Mansilla and Lista both argued that the mixing of races led to the betterment of mankind. In fiction, Holmberg did the same, casting the mixed-race Lin-Calél as the womb of the future Argentine nation. In his poem, he inserted elements of a number of evolutionary theories, including the struggle for life, sexual selection, and Lamarck's soft inheritance, into a romantic framework to shape the Argentine past into a usable future. Eduarda Mansilla also saw racial mixing as the base of the Argentine nation in both *Lucía Miranda* and her later travelogue *Recuerdos de viaje*. Although concerned about masculine indigenous sexuality, she argued that appropriate interracial pairings could found the nation. Echoing concerns of racial scientists around the world, including her compatriot Lista, Eduarda Mansilla argued that it was important that the representatives of each race also be of the right gender. Indigenous women could mix with Spanish men, but the reverse was never a good idea.

Together, the focus on gender and sex and the interdisciplinary nature of my analysis work to question the historiographic tradition that tends to interpret nineteenth-century Argentine science and its effects as either a heroic moment of nation building or a despicable moment of genocide. As this book has shown, it was often both and sometimes neither. Concerns about gender and sex underscored many of the racist characterizations of indigenous peoples and justified their destruction, but other concerns about gender, particularly with regard to Christian motherhood, raised the public's hackles and led to policy changes. Scientists changed their minds, adopting a more positive assessment of indigenous peoples (Lista) and more negative ones (Lucio Mansilla), sometimes because of their own experiences of desire. Moreno managed to view Inacayal, Foyel, and their families as simultaneously friends, citizens in training, and scientific specimens perfect for displaying in the comparative anatomy galleries. For every text by Zeballos arguing for extermination, there was another by Lista or Holmberg or Dixie or a Mansilla arguing for incorporation. Scientific racism was an important and often destructive force, but its consequences were not inevitable. Unlike the perceived amorality of evolution, even during the nineteenth century numerous people recognized Argentine actions as cruel.

It is also important to remember that late nineteenth-century racial science was not isolated in time. Eduardo Holmberg, Juan Zorrilla de San Martín, and Eduarda Mansilla set their fictional explorations of race in the colonial period prior to the foundation of Argentina as nation. Their examination of racial roots prior to Argentine history proper literarily echoed the anthropologists' search for the fossil bones and artifacts that could establish a scientific racial prehistory for the nation. In similar fashion, many of the racial definitions used in the literary and scientific texts of the nineteenth century can be traced back to the colonial period, if not before.

More importantly, we can trace the legacy of these scientific processes in the 150 years that have followed. For many, Argentina's indigenous history appears to have ended in the late 1880s. Nineteenth-century Argentines declared the Conquest of the Desert a rousing success. In 1885, General Lorenzo Vintter announced that Argentina now had total control over all indigenous peoples in the southern countryside.[1] Four years later, in 1889, French engineer Alfred Ebelot proclaimed that in Argentina, "The Indian no longer exists."[2] Similarly, the vast majority of historical, literary, and cultural studies of Argentina follow the example of Vintter and Ebelot and assume a firm break around 1885, entirely separating the history of the Indian wars from the immigration and eugenics movements of the twentieth century. Nonetheless, indigenous people did not actually disappear at the end of the nineteenth century. Rather, new definitions (indios become pobres or campesinos), shifting focuses (immigration), and the temporally othering processes of anthropology served to hide the fact that indigenous peoples were still alive.

The tangle of sex, women, and race that I have traced throughout this book similarly endured far into the future. In many aspects, the experience of assimilating natives was a proving ground for the ways of thinking that would later mediate the incorporation of immigrants and the bettering of the race through eugenics. Practically, the two moments were closely tied together, as the mid-nineteenth-century desire to whiten the nation by populating indigenous lands with civilized bodies led to the permissive immigration policies and propaganda that attracted immigrants.

The ties were also theoretical, as the scientific theories developed around the *cuestión de indios* paved the way for the adoption of eugenic and criminological policy beginning around 1910. The similarities between Moreno's portraits of the Tehuelches and Bertillon-style photography are so great that they have led many scholars to posit that Moreno was aware of bertillonage, despite the fact that it could not have yet arrived in Argentina. The phrenological and anthropometrical study of indigenous bodies for signs of racial characteristics similarly anticipated the work of Cesare

Lombroso. These connections can be seen clearly in the figure of Clemente Onelli. In the early 1880s, the Italian immigrant Clemente Onelli travelled with Moreno, searching for bones and fossils to try to clarify the prehistorical origins of mankind and thus the relative value of the white and indigenous races in the River Plate. Twenty years later, he sent a Tehuelche skull to Lombroso in order to contribute to his comparisons between criminal types and the atavistic primitivism of native people.[3] Onelli also applied Lombroso's work on criminality and alcoholism to his depiction of the Argentine Indians, whom he saw as victims not of inferior biology, but of alcohol and poor hygiene.[4]

Focusing on sex and women makes the bridges between the *cuestión de indios* and immigrant eugenics come into focus. First, as Melissa N. Stein argues for the United States, the focus on sex at the turn of the century was "an extension of earlier developments . . . gender *was always* integral to scientific understanding of race from the start, but the function it served changed over time with shifts in the nation's cultural and political landscape."[5] In Argentina, nineteenth-century theoretical concerns about evolution, degeneration, and hybridity (as a result of sex) as pertaining to indigenous populations set the vocabulary and the general patterns for later applications to burgeoning immigrant groups.

Whereas Zorrilla de San Martín used degeneration and the negative effects of mestizaje to explain the necessary disappearance of the Charrúa Indians, early twentieth-century thinkers worried about degeneration and sexual deviance in relation to poor immigrants and stressed the racial differences between northern and southern Europeans.[6] For example, in 1912, Lucas Ayarragaray published an article in José Ingenieros's *Archivos de psiquiatría y criminología aplicadas* in which he fretted about the degeneration he saw threatening the country, particularly as a result of racial mixing. The lack of care in choosing immigrants, he argued, had allowed the entry of "all the residues of old and exhausted races, which united with the indigenous and mestizo population, have formed truly deplorable ethnic conditions."[7] Immigrants no longer had the luster they did for late nineteenth-century figures like Lista, but sex and its consequences remained at the core of scientific approaches to understanding national identity and possibility.

There are also parallels between the practical manifestations of these concerns before and after the turn of the century. Nancy Stepan has argued that "Gender was important to eugenics because it was through sexual reproduction that the modification and transmission of the hereditary makeup of future generations occurred. Control of that reproduction, by direct or indirect means, therefore became an important aspect of all eugenics movements."[8] But again, efforts to control reproduction did not emerge uniquely in the early twentieth century. The imprisonment of indigenous peoples like Inacayal, Foyel, and their families foreshadowed the confinement of criminals, prostitutes, and degenerates, removing them from the pool of sexual partners.[9] The distribution policies of the 1870s and early 1880s similarly focused on controlling sexual relationships and limiting certain types of reproduction, thus dictating the racial makeup of the nation. Finally, fictional texts like *Painé*, *Tabaré*, and *Lin-Calél* used murder and suicide to prohibit improper sexual relationships, translating racial recommendations for a lay audience. Although nineteenth-century priorities and policies were different, Argentine elites were already zeroing in on sex and reproduction as key areas for shaping the nation's future.

Finally, the nineteenth century's focus on women as a fundamental part of these processes paved the way for the maternalist spirit of later decades. As we have seen, racial scientists and writers like Moreno, Holmberg, Zorrilla de San Martín, and Eduarda Mansilla argued that indigenous women could more easily be civilized. They also highlighted women's role in cultural reproduction, demonstrating their great impact through their ability to educate future generations. Similarly, the hygienists and eugenicists of the twentieth century argued that women's bodies were most easily transformed due to their passivity. They also redefined women's generational influence in biological terms, arguing that their wombs were the most impactful and immediate "environment" for future populations.[10]

Begun in the mid-1880s and finally published in 1910, Holmberg's poem *Lin-Calél* is a clear manifestation of the connections between the two time periods. One on hand, the poem focuses on indigenous people and their role in national identity. It returns to colonial times, just prior to independence, to ethnographically and poetically recover indigenous peoples already invisible at the moment it appeared in print. On the other hand, in passages relating to the new, Argentine race that was coming, Holmberg anticipates the focus on maternalism and racial improvement that would come to define Argentine eugenics in the decades after his poem's publication:

> A seed of courage and vigor,
> A type in gestation which in its softness
> Will become the masses,
> And docile to the model offered to it,
> Shall sink into the bosom of libertines,
> Or rise to the summits of Glory
> On the winds of atavistic virtue.
> Here's to you, seed of the uncertain future! (306–8)

Here, the insistence on the seed in "gestation" points to the fundamental importance of the mother, while the "model offered to it" suggests the positive changes of eugenics not through killing or impeding of reproduction, but through the improvement (via education, hygiene, etc.) of the progeny. *Lin-Calél* thus clearly represents how the anthropological foci of the nineteenth century ideologically overlapped with eugenic policies, facilitating their incorporation into society in the twentieth century. Further investigation will clarify these mechanisms, points of similarity and difference, and whether or not eugenic practices were used on the indigenous groups made invisible after the Conquest of the Desert.

While popular and academic thinking often suggested that indigenous peoples had been exterminated by the end of the nineteenth century, this past would not be erased forever. Almost exactly one hundred years after the Conquest of the Desert, scientific racism returned to national consciousness through debates over the fate of the indigenous bones that had languished in the Museo de La Plata's basement since they were removed from display in the 1940s, including those of Inacayal's wife.

Historian José Mayo made the first claim for indigenous remains in 1973 and other claims followed in the early 1990s. The successful resolution of these projects was impeded by the fact that the Argentine Civil Code established that materials in the museum's collections formed part of the public domain. Restitution would require a law reclassifying them.[11] Such a law was passed in May 1991. However, it was not until 1994 that the museum released Inacayal's remains and he was returned to Tecka. The same year, reforms to the Argentine Constitution formally recognized indigenous peoples' rights, opening the door for progress on claims to other remains.[12] Hundreds of indigenous people have been returned to their communities of origin from museums around the country since then. Argentine news outlets have reported quite extensively on the claims and various reburial ceremonies, raising awareness among the public about Argentina's troubled past.

Scholars and journalists often tie the experience of nineteenth-century indigenous peoples to that of the disappeared during the 1976–1983 military dictatorship. One example is the traveling photographic exhibition *Prisioneros de la ciencia* (Prisoners of science), curated by Colectivo GUIAS. GUIAS has twice mounted the exhibit, which includes the Boote photographs of the Tehuelches I study in Chapter Three, in the former Escuela Superior de Mecánica de la Armada (ESMA, a military school). During the period of state terrorism, the ESMA was one of the largest clandestine detention and torture centers in the country; the presence of the Boote photographs within this space clearly resituates them within the context of human rights violations.

The exhibit also demonstrates that the debate over restitution and the value of nineteenth-century science has, like its nineteenth-century predecessors, spread into popular culture. Most interestingly from the perspective of sex and gender, the first two decades of the twenty-first century saw a surge in historical romance novels set in the nineteenth century. Many engaged critically with the anthropology of the time. Most representative of this trend are Florencia Bonelli's novels *Indias blancas* and *Indias blancas: La vuelta del ranquel*, and Gloria Casañas's *La maestra de la laguna*, all bestsellers.[13] The novels attempt to reassess Argentina's indigenous heritage, incorporating it into the symbolic national family. Each features a mestizo romantic hero who falls in love with a woman of European descent, and the happy endings suggest this blend was a positive one for the nation. In each, there is a villainous, grave-robbing scientist who threatens the happiness of the protagonists. Bonelli and Casañas also reanimate major figures of nineteenth-century science, including both Mansillas, Sarmiento, and Zeballos. The debates over restitution create an epilogue to Bonelli's second novel, with the return of Chieftain Mariano Rosas' bones from the Museo de La Plata to his community featuring as a second happy ending. While both novels aim to shine light on the racism of nineteenth-century science and celebrate the nation's indigenous and mestizo heritage, I have argued elsewhere that plot choices and the conventions of the romance genre undermine these projects and highlight the staying power of nineteenth-century paradigms.[14]

In addition to Bonelli and Casañas' novels, in 2012, Myriam Angueira and Guillermo Glass released the documentary *Inacayal: La negación de nuestra identidad* (Inacayal: The denial of our identity), that tells Inacayal's story from the nineteenth century until his restitution.[15] The figures of Eduarda and Lucio V. Mansilla have also reappeared in numerous other forms, some critical and some celebratory. In 2017, a group of Argentine universities worked together to create an eight-part series entitled *Otra excursión a los indios ranqueles* (Another visit to the Ranquel Indians). Simultaneously debuted on twenty-four university channels across the country, the program combined fictional reenactment with documentary storytelling and interviews with historians, Ranquel Indians, and other figures related to Lucio Mansilla's narration. This was not the first recreation of Mansilla's excursion, however, for in 1992 Argentine historian and fiction writer María Rosa Lojo retraced his steps on horseback.[16] She wrote about her experience and used it as inspiration for a historical-fantasy novel featuring Mansilla, *La pasión de los nómades*, later translated to English as *Passionate Nomads*.[17]

Lojo has brought other nineteenth-century scientists and intellectuals to the present, publishing *Una mujer de fin de siglo*, historical fiction about Eduarda Mansilla, and *Historias ocultas en la Recoleta*, a series of short stories that includes two about Ramón Lista, his wife Agustina Andrade, and the scandal that led to her suicide.[18] Other works of historical fiction have focused on the Fuegian Indians taken to England who returned to Patagonia with Darwin on the HMS Beagle and on Moreno's collaborator, Clemente Onelli.[19] As each of these texts' authors attempts to challenge or update the legacy of Argentine racial science, they also bring contemporary understandings of sex and gender to their writing. Many are explicitly feminist. How are these interpretations of the racial projects of the nineteenth century transformed by coming into contact with present-day paradigms relating to gender roles, sexual relations, and reproduction?

On December 9, 2014, authorities at the Museo de La Plata turned over the remains of the cacique Inacayal's wife and those of Margarita Foyel to representatives of the Mapuche-Tehuelche community of Chubut (Patagonia).[20] For over 125 years, their bodies had served scientific purposes, involuntarily cast in a charade designed to reinforce white men's power over the Argentine territory and its inhabitants. As observed subjects in Tecka, informants and photographic subjects in Retiro, and skeletons in La Plata, they were undeniably important to the construction of scientific knowledge. Although time and biases have meant that most of the details of their lives have been forgotten or erased, it is my hope that this book is a first step to recognizing their personhood, experiences, and sacrifices, as well as those of the other women who found themselves in nineteenth-century Argentine racial science's path.

Notes

INTRODUCTION

1. Francisco P. Moreno, *Viaje a la Patagonia austral, emprendido bajo los auspicios del Gobierno Nacional, 1876–1877* (Buenos Aires: Imp. de la Nación, 1879), 105.
2. Ramón Lista, *Viaje al país de los tehuelches*, in *Obras*, ed. Jorge Carman (Buenos Aires: Editorial Confluencia, 1998), 1:88.
3. Estanislao Zeballos, *Descripción amena de la República Argentina. Tomo 1. / Viaje al país de los Araucanos* (Buenos Aires: Jacobo Peúser, 1881), 75.
4. John Charles Chasteen, introduction to *Beyond Imagined Communities: Reading and Writing the Nation in Nineteenth-Century Latin America*, ed. Sara Castro-Klarén and John Charles Chasteen (Washington, DC: Woodrow Wilson Center Press, 2003), xviii.
5. Nancy P. Appelbaum, Anne S. Macpherson, and Karin Alejandra Rosemblatt, introduction to *Race and Nation in Modern Latin America*, ed. Appelbaum, Macpherson, and Rosemblatt (Chapel Hill: University of North Carolina Press, 2003), 2.
6. Juan M. Vitulli and David M. Solodkow, "Ritmos diversos y secuencias plurales: Hacia una periodización del concepto 'criollo,'" in *Poéticas de lo criollo: La transformación del concepto "criollo" en las letras hispanoamericanas (siglo XVI al XIX)*, ed. Juan M. Vitulli and David M. Solodkow (Buenos Aires: Corregidor, 2009), 11. Throughout Latin American history, the identifier *Creole* (*criollo*) has been constantly transformed and given new significance. During the early colonial period, Creoles were people of Spanish descent born in the Americas, distinguishing them from the Peninsulars (*peninsulares*) born in Spain. In the late-sixteenth and seventeenth centuries, those born in Spanish America had taken control of the term and started to use it to positively distinguish themselves from Spain. *Creole*'s weight as an identifier and unifying force increased over time, reaching a peak during the in-

dependence wars in the early nineteenth century. Racially, *Creole* existed primarily as a marker of difference. In the face of indigenous peoples, it continued to mean those of European descent. When contrasting with immigrants or European powers, it could refer to all Argentines, even those of mixed race. In other cases, it meant those from the interior as opposed to those in Buenos Aires. I generally use *Creole* to refer to those Argentines who saw themselves as not indigenous.

It is worth noting that people of African descent also played an important role in the everyday life and nation-building processes in Argentina. Nonetheless, they were almost completely absent from the anthropological discourse of the time. On Afro-Argentines, see George Reid Andrews, *The Afro-Argentines of Buenos Aires, 1800–1900* (Madison: University of Wisconsin Press, 1980); Silvia C. Mallo and Ignacio Telesa, eds., *"Negros de la patria": Afrodescendientes en las luchas por la independencia en el antiguo virreinato del Río de la Plata* (Buenos Aires: Editorial SB, 2010); Ricardo D. Salvatore, "Integral Outsiders: Afro-Argentines in the Era of Juan Manuel de Rosas and Beyond," in *Beyond Slavery: The Multilayered Legacy of Africans in Latin America and the Caribbean*, ed. Darién J. Davis (Lanham: Rowan & Littlefield, 2007): 57–80.

7. María E. Argeri, *De guerreros a delincuentes: La desarticulación de las jefaturas indígenas y el poder judicial. Norpatagonia, 1880–1930* (Madrid: Consejo Superior de Investigaciones Científicas, 2005), 14.

8. Lista, *Viaje al país de los tehuelches*, 88.

9. Rebecca K. Jager, *Malinche, Pocahontas, and Sacagawea: Indian Women as Cultural Intermediaries and National Symbols* (Norman: University of Oklahoma Press, 2015), 3.

10. George Chaworth Musters, *At Home with the Patagonians*, 2nd ed. (London: John Murray, 1873), 99, *babel.hathitrust.org/cgi/pt?id=mdp.39015018047343*.

11. Carlos Martínez Sarasola, *Nuestros paisanos los indios: Vida, historia y destino de las comunidades indígenas en la Argentina* (Buenos Aires: Del Nuevo Extremo, 2013), 393.

12. For example, Ramón Lista wrote General Roca during the Conquest of the Desert, asking him to gather Araucanian skulls that Lista could use to finish an article on prehistoric times. Ramón Lista to Julio A. Roca, June 1879, sala VII legajo 7, Julio Argentino Roca archive, Archivo General de la Nación, Buenos Aires, Argentina.

13. Scholars like Marcelo Valko, Diana Lenton, Walter Delrio, and the Colectivo GUIAS re-envision national heroes like Moreno as villains and accuse once-celebrated scientific institutions such as the Museo de La Plata as complicit in a mass genocide. Walter Delrio et al., "Discussing Indigenous Genocide in Argentina: Past, Present, and Consequences of Argentinean State Policies toward Native Peoples," *Genocide Studies and Prevention: An International Journal* 5, no. 2 (2010): 138–59; Diana Isabel Lenton, "Relaciones interétnicas: Derechos humanos y autocrítica en la Generación Del '80," in *La problemática indígena: Estudios antropológicos sobre pueblos indígenas de la Argentina*, ed. Juan Carlos Radovich and Alejandro O. Balazote (Buenos Aires: Centro Editor de América Latina, 1992), 27–65; Marcelo Valko, *Pedagogía de la desmemoria: Crónicas y estrategias del genocidio invisible* (Bue-

nos Aires: Continente, 2013); and Marcelo Valko, *Cazadores de poder: Apropiadores de indios y tierras (1880–1890)* (Buenos Aires: Ediciones Continente, 2015). The Colectivo GUIAS is a group of anthropologists and sociologists based out of La Plata, Argentina, who have worked to identify and facilitate the return of indigenous remains once displayed in the Museo de La Plata. They have published numerous books of photography that denounce the actions of nineteenth-century scientists, including Fernando Miguel Pepe, Miguel Añon Suarez, and Patricio Harrison, *Antropología del genocidio: Identificación y restitución. "Colecciones" de restos humanos en el Museo de La Plata* (La Plata: De la Campana, 2010); and Fernando Miguel Pepe et al., *"Bioiconografía": Los prisioneros de la "Campaña del desierto" en el Museo de La Plata, 1886* (La Plata: De la Campana, 2013). These interventions have been fruitful and necessary, but they have also muted some of the complexity of the period.

14. "Indigenous Peoples in Latin America," infographic, CEPAL, September 22, 2014, *www.cepal.org/en/infografias/los-pueblos-indigenas-en-america-latina.*

15. Mary Louise Pratt, *Imperial Eyes: Travel Writing and Transculturation* (New York: Routledge, 1992), 4.

16. Nora Siegrist de Gentile and María Haydée Martín, *Geopolítica, diencia y técnica a través de la campaña del desierto* (Buenos Aires: Editorial Universitaria de Buenos Aires, 1981), 49.

17. Enrique Hugo Mases, *Estado y cuestión indígena: El destino final de los indios sometidos en el sur del territorio (1878–1930)* (Buenos Aires: Prometeo, 2010), 30. For more on reciprocal practices of captivity, see Fernando Operé, *Indian Captivity in Spanish America: Frontier Narratives*, trans. Gustavo Pellón (Charlottesville: University of Virginia Press, 2008); and Susana Rotker, *Captive Women: Oblivion and Memory in Argentina*, trans. Jennifer French (Minneapolis: University of Minnesota Press, 2002).

18. Mónica Quijada, "La ciudadanización del 'indio bárbaro': Políticas oficiales y oficiosas hacia la población indígena de la Pampa y la Patagonia, 1870–1920," *Revista de Indias* 109, no. 217 (1999): 676–78.

19. On indigenous participation in Argentine defenses against the English invasions, see Pablo Andrés Cuadra Centeno and María Laura Mazzoni, "La invasión inglesa y la participación popular en la reconquista y defensa de Buenos Aires 1806–1807," *Anuario del Instituto de Historia Argentina*, no. 11 (2011): 55–57. On Argentine-indigenous cultural, economic, political, and biological integration, see Mónica Quijada, "La ciudadanización del 'indio bárbaro'"; Mónica Quijada, "Repensando la frontera sur argentina: Concepto, contenido, continuidades y discontinuidades de una realidad espacial y étnica (siglos XVIII–XIX)," *Revista de Indias* 62, no. 224 (2002): 103–42; Raúl Mandrini, "Indios y fronteras en el área pampeana (siglos XVI–XIX). Balance y perspectivas," *Anuario del IEHS* 7 (1992): 59–72; Raúl Mandrini and Andrea Reguera, *Huellas en la tierra: Indios, agricultores y hacendados en la pampa bonaerense* (Tandil: IEHS, 1993); Lidia R. Nancuzzi, "'Nómades' versus 'sedentarios' en Patagonia (siglos XVIII-XIX)," *Cuadernos del Instituto Nacional de Antropología y Pensamiento Latinoamericano* 14 (1992–93): 81–92.

20. Martínez Sarasola, *Nuestros paisanos*, 381.

21. On the Generation of '80 see Hebe Noemí Campanella, *La generación del 80: Su influencia en la vida cultural argentina* (Buenos Aires: Tekné, 1983); Hugo Biagini, *Cómo fue la generación del 80* (Buenos Aires: Plus Ultra, 1980); and Bonnie Frederick, "In Their Own Voice: The Women Writers of the Generación del 80 in Argentina," *Hispania* 74, no. 2 (May 1991): 282–89.

22. Ruth Hill, "Entre lo transatlántico y lo hemisférico: Los proyectos raciales de Andrés Bello," *Revista Iberoamericana* 75, no. 228 (2009): 723. On the weakness of the Americas, see Antonello Gerbi, *The Dispute of the New World: The History of a Polemic, 1750–1900*, trans. Jeremy Moyle (Pittsburgh: University of Pittsburgh Press, 2010).

23. Domingo F. Sarmiento, *Conflicto y armonías de las razas en América* (Buenos Aires: Ostwald, 1883), 1, *catalog.hathitrust.org/Record/000279974*.

24. Manuel Ricardo Trelles, "Al señor ministro de gobierno Doctor Don Nicolás Avellaneda," *La revista de Buenos Aires* 16 (1868): 590.

25. José Babini, *La ciencia en la Argentina*, (Buenos Aires: Eudeba, 1963), 41–76; Miguel de Asúa, *Una gloria silenciosa: Dos siglos de ciencia en la Argentina* (Buenos Aires: Libros del Zorzal, 2010); Marcelo Monserrat, ed. *La ciencia en la Argentina entre siglos: Textos, contextos e instituciones* (Buenos Aires: Manantial, 2000), especially the section "La ciencia en sus instituciones," 277–365.

26. Although today we view anthropology and ethnography as separate fields, the nineteenth-century Argentines I study did not distinguish between them. At various points in their narratives, they call themselves anthropologists and ethnographers, as well as explorers and scientists. For ease of reading, I have chosen to refer to them primarily as anthropologists or racial scientists.

27. For most of the nineteenth century the concept of the Argentine nation was almost exclusively based on territory. It was this formulation that allowed indigenous peoples to be classified as Argentines according to the 1853 Constitution, which allocated equal rights to all inhabitants of the territory (articles 14, 16). Argentine Republic, "Constitución de la Confederación Argentina (May 1, 1853)," WIPO Lex No. AR147, *www.wipo.int/wipolex/en/text.jsp?file_id=361080*.

28. Mases, *Estado y cuestión indígena*, 55–56.

29. Martínez Sarasola, *Nuestros paisanos*, 278.

30. Qtd. in Mases, *Estado y cuestión indígena*, 59.

31. Pablo Azar, Gabriela Nacach, and Pedro Navarro Floria, "Antropología, genocidio y olvido en la representación del otro étnico a partir de la Conquista," in *Paisajes del progreso: La resignificación de la Patagonia Norte, 1880–1916*, ed. Pedro Navarro Floria (Neuquen: Educo, 2007), 79–106; Pedro Navarro Floria, "El salvaje y su tratamiento en el discurso político argentino sobre la frontera sur, 1853–1879," *Revista de Indias* 61, no. 222 (2001): 345–76; and Mónica Quijada, "¿'Hijos de los barcos' o diversidad invisibilizada? La articulación de la población indígena en la construcción nacional argentina (siglo XIX)," *Historia mexicana* 53, no. 2 (Oct.–Dec. 2003): 469–510.

32. Ruth Hill, "Ariana Crosses the Atlantic: An Archaeology of Aryanism in the Nineteenth-Century River Plate," *Hispanic Issues On Line* 12 (2013): 92–110, *hdl.handle.net/11299/184420*; Ashley Elizabeth Kerr, "'Somos una raza privilegiada': Anthropology, Race, and Nation in the Literature of the River Plate, 1870–2010" (PhD diss., University of Virginia, 2013); Pedro Navarro Floria, Leonardo Salgado, and Pablo Azar, "La invención de los ancestros: El 'patagón antiguo' y la construcción discursiva de un pasado nacional remoto para la Argentina (1870–1915)," *Revista de Indias* 64, no. 231 (2004): 405–24; and Mónica Quijada, "Los 'incas arios': Historia, lengua y raza en la construcción nacional hispanoamericana del siglo XIX," *Histórica* 20, no. 2 (1996): 243–69.

33. Adriana Novoa and Alex Levine, *From Man to Ape: Darwinism in Argentina, 1870–1920* (Chicago: University of Chicago Press, 2014), 122; Adriana Novoa, "The Rise and Fall of Spencer's Evolutionary Ideas in Argentina, 1870–1910," in *Global Spencerism: The Communication and Appropriation of a British Evolutionist*, ed. Bernard Lightman (Leiden: Brill, 2015), 182.

34. Máximo Farro, *La formación del Museo de La Plata: Coleccionistas, comerciantes, estudiosos y naturalistas viajeros a fines del siglo XIX* (Rosario: Prohistoria, 2009), 17. Farro, María Margaret Lopes, and Irina Podgorny's archival work has revealed much of the museum's history, while Jens Andermann, Carolyne R. Larson, and Mónica Quijada have looked at the connections between museum culture, indigenous peoples, and national identity. Jens Andermann, *The Optic of the State: Visuality and Power in Argentina and Brazil* (Pittsburgh: University of Pittsburgh Press, 2007); Carolyne R. Larson, *Our Indigenous Ancestors: A Cultural History of Museums, Science, and Identity in Argentina, 1877–1943* (University Park: Pennsylvania State University Press, 2015); Irina Podgorny, *El sendero del tiempo y de las causas accidentales: Los espacios de la prehistoria en la Argentina, 1850–1910* (Rosario: Prohistoria, 2009); Irina Podgorny and María Margaret Lopes, *El desierto en una vitrina: Museos e historia natural en la Argentina, 1810–1890* (México: Limusa, 2008); and Mónica Quijada, "Ancestros, ciudadanos, piezas de museo: Francisco P. Moreno y la articulación del indígena en la construcción nacional argentina (siglo XIX)," *Estudios interdisciplinarios de América Latina y el Caribe* 9, no. 2 (1998): 21–46.

35. Larson, *Our Indigenous Ancestors*, 4.

36. Azar, Nacach, and Navarro Floria, "Antropología, genocidio y olvido," 79; and Jens Andermann, *Mapas de poder: Una arqueología literaria del espacio argentino* (Rosario: Beatriz Viterbo, 2000), 125.

37. One notable exception is Novoa and Levine's *From Man to Ape*, which includes a chaper on sexual selection and Argentine understandings of gender.

38. See, for example, Elizabeth Thompson's *The Pioneer Woman: A Canadian Character Type* (Montreal: McGill-Queen's University Press, 1991) and Ann Laura Stoler, *Carnal Knowledge and Imperial Power: Race and the Intimate in Colonial Rule* (Berkeley: University of California Press, 2002). For studies on these relations in Argentina specifically, see Bonnie Frederick, *Wily Modesty: Argentine Women Writers, 1860–1910* (Tempe: ASU Center for Latin American Studies Press, 1998); Francine

Masiello, *Between Civilization and Barbarism: Women, Nation and Literary Culture in Modern Argentina* (Lincoln: University of Nebraska Press, 1992); Nancy Hanway, *Embodying Argentina: Body, Space and Nation in 19th Century Narrative* (Jefferson: McFarland & Co., 2003); and Donna Guy, *Sex and Danger in Buenos Aires: Prostitution, Family, and Nation in Argentina* (Lincoln: University of Nebraska Press, 1991).

39. Nancy Stepan, *"The Hour of Eugenics": Race, Gender, and Nation in Latin America* (Ithaca, NY: Cornell University Press, 1991); Julia Rodríguez, *Civilizing Argentina: Science, Medicine, and the Modern State* (Chapel Hill: University of North Carolina Press, 2006). Outside of Latin America, Siobhan Somerville's manuscript examining race and the rise of homosexuality (2000) and Melissa Stein's 2015 study of the role in gender in making and unmaking race in the United States do look at the nineteenth century. Siobhan Somerville, "Scientific Racism and the Emergence of the Homosexual Body," *Journal of the History of Sexuality* 5, no. 2 (October 1994): 243–66; Melissa N. Stein, *Measuring Manhood: Race and the Science of Masculinity, 1830–1934* (Minneapolis: University of Minnesota Press, 2015).

40. Ann Laura Stoler, *Carnal Knowledge and Imperial Power*; Robert Young, *Colonial Desire: Hybridity in Theory, Culture and Race* (London: Routledge, 1995); and Sander L. Gilman, *Difference and Pathology: Stereotypes of Sexuality, Race, and Madness* (Ithaca, NY: Cornell University Press, 1985).

41. Peter Wade, *Race and Sex in Latin America* (London: Pluto Press, 2009), 114.

42. For a discussion of this relationship in present-day anthropological texts, see Esther Newton, "My Best Informant's Dress: The Erotic Equation in Fieldwork," *Cultural Anthropology* 8, no. 1 (February 1993): 4.

43. Silvia Hirsch, "La mujer indígena en la antropología argentina: Una breve reseña," in *Mujeres indígenas en la Argentina: Cuerpo, trabajo y poder*, ed. Silvia Hirsch (Buenos Aires: Biblos, 2008), 17.

44. Andermann, *The Optic of the State.*

45. Sujit Sivasundaram, "Sciences and the Global: On Methods, Questions, and Theory," *Isis* 101, no. 1 (March 2010): 146, 154. See also James Secord, "Knowledge in Transit," *Isis* 95, no. 4 (December 2004): 654–72.

46. Anna Brickhouse, *The Unsettlement of America: Translation, Interpretation, and the Story of Don Luis de Velasco, 1560–1945* (Oxford: Oxford University Press, 2015), 19.

47. David Viñas, *Indios, ejército y frontera* (Mexico: Siglo Veintiuno, 1982), 227.

48. Mariela Eva Rodríguez, "De la 'extinción' a la autoafirmación: Procesos de visibilización de la comunidad tehuelche Camusu Aike (Provincia de Santa Cruz, Argentina)" (PhD diss., Georgetown University, 2010), 68, *repository.library.georgetown.edu/bitstream/handle/10822/553246/rodriguezMariela.pdf.*

49. Rodolfo Casamiquela defended Moreno, arguing that he had been unfairly blacklisted and offering as proof the fact that other indigenous peoples later came to the museum as free visitors. In *Cazadores de poder*, Marcelo Valko skewers Moreno, claiming that he "enjoyed like no other being surrounded by the skulls of the defeated." Of particular importance to this current is the work of Colectivo

GUIAS, a group of sociologists and anthropologists working to return indigenous remains in the Museo de La Plata to their communities of origin. In their very successful photographic exposition and two books, these scholar-activists refer to the Tehuelches and others as "prisoners of science" and speak of an "anthropology of genocide," two terms that highlight self-interested scientific motivations while casting doubt on Moreno's claim to humanitarianism. Rodolfo Casamiquela, "Indígenas patagónicos en el museo: Primera parte," *Revista museo* 11, no. 12 (1998): 69–75; Valko, *Cazadores de poder*, 16; Colectivo GUIAS: Grupo Universitario de Investigación en Antropología Social, home page, accessed April 29, 2018, *colectivoguias.blogspot.com*.

CHAPTER 1

1. Juan Batista Alberdi, *Bases y puntos de partida para la organización política de la República Argentina* (Buenos Aires: La Cultura Argentina, 1928), 14, *archive.org/download/basesypuntosdepa00albe/basesypuntosdepa00albe.pdf*; José Mármol, *Amalia*, ed. Teodosio Fernández (Madrid: Cátedra, 2010).
2. The texts I review include Ramón Lista's *Viaje al país de los tehuelches* (Journey to the land of the Tehuelches, 1879), *El Territorio de las Misiones* (The territory of Misiones, 1883), and *Viaje al país de los onas* (Journey to the land of the Onas, 1887); Lucio V. Mansilla's *Una excursión a los indios ranqueles* (1870; published in translation as *A Visit to the Ranquel Indians*, 1997); Francisco P. Moreno's *Viaje a la Patagonia austral* (Journey to southern Patagonia, 1879); and Estanislao Zeballos's *La conquista de quince mil leguas* (The conquest of fifteen thousand leagues, 1878), *Descripción amena de la República Argentina* (Pleasant description of the Argentine Republic, 1881), *Callvucurá y la dinastía de los Piedra* (Callvucurá and the Piedra dynasty, 1884), *Painé y la dinastía de los Zorros* (Painé and the Zorro dynasty, 1886), and *Relmú, reina de los pinares* (Relmú, queen of the pine forests, 1888).
3. John Lubbock, *The Origin of Civilisation and the Primitive Condition of Man: Mental and Social Condition of Savages* (New York: D. Appleton and Co., 1870), 50, *catalog.hathitrust.org/Record/100265213*.
4. Elizabeth Potter, *Feminism and Philosophy of Science: An Introduction* (London: Routledge, 2006), 60. Potter summarizes the work of Clifford Geertz and discusses a variety of later revisions to it.
5. In recent years, scholars have begun to publish on the existence of nonbinary ideas of gender among Southern Cone indigenous groups. For example, Bacigalupo discusses the cogendered *machi weye* tradition among the Mapuche in Chile during the colonial and republican periods. Given Mapuche influence on Argentina during this time (indeed, the anthropologists spoke frequently of the "araucanization" of the region), it is probable that similar gender structures existed. Other scholars such as Pablo Ben, Cristian Berco, and Jorge Salessi have looked at the practice and prohibition of same-sex relationships in the republican period. Neverthless, neither nonbinary gender expressions nor same-sex relationships are ever mentioned by the men that I study. Thus, while recognizing that they were likely im-

portant, I exclude them from my analysis. Ana Mariella Bacigalupo, *Shamans of the Foye Tree: Gender, Power, and Healing among Chilean Mapuche* (Austin: University of Texas Press, 2007), 111–34; Pablo Ben, "Male Same-Sex Sexuality and the Argentine State, 1880–1930," in *The Politics of Sexuality in Latin America*, ed. Javier Corrales and Mario Pecheny (Pittsburgh: University of Pittsburgh Press, 2010), 33–43; Cristian Berco, "Silencing the Unmentionable: Non-Reproductive Sex and the Creation of a Civilized Argentina, 1860–1900," *The Americas* 58, no. 3 (2002): 419–41; Jorge Salessi, "The Argentine Dissemination of Homosexuality, 1890–1914," *Journal of the History of Sexuality* 4, no. 3 (1994): 337–68.

6. For a discussion of these shifts, see Adriana Novoa, "Unclaimed Fright: Race, Masculinity, and National Identity in Argentina, 1850–1910," PhD diss, University of California, San Diego, 1998.

7. Frederick, *Wily Modesty*, 85.

8. Marcela Nari, *Políticas de maternidad y maternalismo politico: Buenos Aires, 1890–1940* (Buenos Aires: Biblos, 2004), 63. The reality of women of other classes can be seen in Donna Guy's work including Donna Guy, "Lower-Class Families, Women, and the Law in Nineteenth-Century Argentina," *Journal of Family History* 10, no. 3 (1985): 318–31.

9. Ana Pelufo and Ignacio M. Sánchez Prado, eds., *Entre hombres: Masculinidades del siglo XIX en América Latina* (Madrid: Iberoamericana, 2010), 9. Despite history and cultural criticism's tendency to focus on male actors, there have been relatively few studies of masculinity in nineteenth-century Argentina. For a good overview of existing research, see Carolina Rocha, ed., *Modern Argentine Masculinities* (Bristol: Intellect, 2013). Other key texts on Argentine or Latin American masculinities include Matthew Gutmann, *Changing Men and Masculinities in Latin America* (Durham: Duke University Press, 2002); and Peluffo and Sánchez Prado, eds., *Entre hombres*.

10. Argeri, *De guerreros a delicuentes*, 210–12.

11. Asunción Lavrin, *Women, Feminism, and Social Change in Argentina, Chile, and Uruguay, 1890–1940* (Lincoln: University of Nebraska Press, 1995), 194–95.

12. Donna Guy, "Lower-Class Families, Women, and the Law," 325.

13. Kif Augustine-Adams, "'She Consents Implicitly': Women's Citizenship, Marriage, and Liberal Political Theory in Late-Nineteenth- and Early-Twentieth-Century Argentina," *Journal of Women's History* 13, no. 4 (2002): 10.

14. Masiello, *Between Civilization and Barbarism*, 19.

15. Alberdi, *Bases*, 80.

16. Nari, *Políticas de maternidad*, 31.

17. Nari, *Políticas de maternidad*, 62–63.

18. Stein, *Measuring Manhood*, 30.

19. Martínez Sarasola, *Nuestros paisanos los indios*, 325.

20. Moreno, *Viaje a la Patagonia austral*,11.

21. Lista, *Viaje al país de los tehuelches*, 90.

22. Claudia Torre, *Literatura en tránsito: La narrativa expedicionaria de la Conquista del Desierto* (Buenos Aires: Prometeo, 2010), 96.

23. Lucio V. Mansilla, *A Visit to the Ranquel Indians,* trans. and ed. Eva Gilles (Lincoln: University of Nebraska Press, 1997), 226.
24. Ramón Lista, *El Territorio de las Misiones,* in *Obras,* ed. Jorge Carman (Buenos Aires: Editorial Confluencia, 1998), 1:291.
25. Erin O'Connor describes a similar relationship in Ecuador, where government officials suggested that indigenous peoples "were unable to identify and defend their own interests, whether the matter at hand was protecting their lands from non-Indians or their own persons from excessive alcohol consumption. They therefore needed white-mestizo father figures to safeguard their rights and teach them civilization and self-restraint." *Gender, Indian, Nation: The Contradictions of Making Ecuador, 1830–1925* (Tucson: University of Arizona Press, 2007), 30.
26. Moreno, *Viaje a la Patagonia austral,* 366.
27. "Los indios," *La Prensa,* December 3, 1878.
28. See, for example, Zeballos, *Descripción amena,* 94.
29. Moreno, *Viaje a la Patagonia austral,* 11; Lista, *Viaje al país de los tehuelches,* 1:90.
30. Ramón Lista, *Viaje al país de los onas,* in *Obras,* ed. Jorge Carman (Buenos Aires: Editorial Confluencia, 1998), 2:22.
31. Lista, *Viaje al país de los onas,* 2:22.
32. Lista, *Viaje al país de los onas,* 2:23.
33. Mansilla, *A Visit,* 329. In their observation and judgement of indigenous peoples, the Argentine anthropologists agreed closely with their European and North American counterparts. The French Arthur de Gobineau described the "American savage" as "idle, and dirty, lazily dragging his feet along his uncultivated ground." Henry Rowe Schoolcraft, the prolific ethnologist of North American Indians, characterized the Algonquians as excessively slothful, attributing the trait to "a luxurious effeminacy, produced upon the race under a climate more adverse to personal activity." Arthur de Gobineau, *The Inequalty of Human Races,* trans. Adrian Collins (London: William Heinemann, 1915), 28, *babel.hathitrust.org/cgi/pt?id=msu.31293103445569;* Schoolcraft, *Algic Researches: North American Indian Folktales and Legends,* ed. Curtis M. Hinsley (Mineola, NY: Dover Publications, 1999), xxi. Furthermore, John Lubbock argued that men's laziness meant that "among low races the wife is indeed literally the property of the husband," tasked with hard labor uncompensated by love or affection. *The Origin,* 68. On both sides of the Atlantic, intellectual elites viewed this treatment of women as one of the classic characteristics of societies in a state of barbarism. Scholars have previously attributed these affinities between the Argentine, European, and North American anthropologists to the European and North American sources' effects on Argentine readers, but shared underlying cultural understandings of the responsibilities of each gender also played a part in the parallel rise of these images.
34. Estanislao S. Zeballos, *La conquista de quince mil leguas: Ensayo para la ocupación definitiva de la Patagonia* (1878) (Buenos Aires: Ediciones Continente, 2008), 275.
35. Estanislao S. Zeballos, *Relmú, reina de los pinares,* 2nd ed. (Buenos Aires: Jacobo Peúser, 1894), 246, *books.google.com/books?id=UD9UAAAAMAAJ.*

36. Jager, *Malinche, Pocahontas, and Sacagawea*, 122.

37. Liliana Videla, "María, la cacica de los tehuelches," *Todo es historia* 477 (2007): 28–35.

38. Argeri, *De guerreros a delincuentes*, 227.

39. María Florencia del Castillo Bernal and Liliana E. María Videla, "Estudio comparativo de tres jefaturas femeninas en Patagonia," *Actas Del VI Congreso de Historia Social y Política de La Patagonia Argentino-Chilena* (Trevelín, Argentina, 2009), 17. Christian Gonzalo Quiroga has also examined women's participation in Tehuelche and Manzanero diplomacy in the late nineteenth century. Cristian Gonzalo Quiroga, "Análisis del rol de las mujeres indígenas en los ámbitos de consenso, durante la segunda mitad del siglo XIX en Patagonia. Sugerencias para una nueva interpretación de caso," in *V Jornadas de historia social de la Patagonia*, comp. Walter Delrio et al. (Río Negro: San Carlos de Bariloche, 2014), 105–30, *biblioteca.clacso.edu.ar/Argentina/iidypca-unrn/20171115052701/pdf_108.pdf*.

40. Miguel Angel Palermo, "El revés de la trama: Apuntes sobre el papel económico de la mujer en las sociedades indígenas tradicionales del sur argentino," *Memoria Americana: Cuadernos de Etnohistoria* 3 (1994): 73–80. Florencia Roulet analyzes the role of women in frontier diplomacy in "Mujeres, rehenes y secretarios: Mediadores indígenas en la frontera sur del Río de la Plata durante el período hispánico," *Colonial Latin American Review* 18, no. 3 (December 2009): 303–37.

41. Moreno, *Viaje a la Patagonia austral*, 219-22.

42. Moreno, *Viaje a la Patagonia austral*, 230-32.

43. Moreno, *Viaje a la Patagonia austral*, 366.

44. For example, Lista, "Viaje al país de los onas," 101; Mansilla, *A Visit*, 195.

45. Operé, *Indian Captivity in Spanish America*, 69.

46. Zeballos, *Descripción amena*, 380.

47. Moreno, *Viaje a la Patagonia austral*, 214-15.

48. Fernanda Peñaloza, "On Skulls, Orgies, Virgins and the Making of Patagonia as a National Territory: Francisco Pascasio Moreno's Representations of Indigenous Tribes," BHS 87, no. 4 (2010): 468.

49. Mansilla, *A Visit*, 222.

50. Zeballos, *Descripción amena*, 257.

51. Estanislao S. Zeballos, *Painé y la dinastía de los Zorros*, 2nd ed. (Buenos Aires: Jacobo Peuser, 1889), 70.

52. Zeballos, *Painé*, 64, for example.

53. Zeballos, *Painé*, 161.

54. Zeballos, *Painé*, 84.

55. Mansilla, *A Visit*, 222.

56. Zeballos, *Descripción amena*, 386.

57. Estanislao S. Zeballos, *Callvucurá y la dinastía de los Piedra*, 3rd ed. (Buenos Aires: J. Peuser, 1890), 19.

58. Gabriella Nouzeilles, "El retorno de lo primitivo: Aventura y masculinidad," in *Entre hombres: Masculinidades del siglo XIX en América Latina*, eds. Ana Peluffo and Ignacio M. Sánchez Prado (Madrid: Iberoamericana, 2010), 98.

59. Rotker, *Captive Women*, 8.
60. Rotker, *Captive Women*, 66.
61. Zeballos, *Painé*, 79.
62. Mansilla, *A Visit*, 222.
63. Roy Harvey Pearce, *Savagism and Civilization: A Study of the Indian and the American Mind*, rev. ed., (Berkeley: University of California Press, 1988), 89.
64. John Browning, "Cornelius de Pauw and Exiled Jesuits: The Development of Nationalism in Spanish America," *Eighteenth-Century Studies* 11, no. 3 (1978): 289–307, *www.jstor.org/stable/2738194*.
65. Susan Migden Socolow, "Spanish Captives in Indian Societies: Cultural Contact along the Argentine Frontier, 1600–1835," *Hispanic American Historical Review* 72, no. 1 (1992): 95.
66. Rotker, *Captive Women*, 68. The inflation of fears of sexual assault by racial minority men has a long history in colonial contexts. For example, Ann Laura Stoler has shown that in much of the early twentieth-century British Empire, including Rhodesia, Kenya, New Guinea, and the Solomon Islands, "the proliferation of discourse about sexual assault and the measures used to prevent it had virtually no correlation with actual incidences of rape of European women by men of color." Stoler, *Carnal Knowledge*, 58.
67. María Bjerg, "Vínculos mestizos: Historias de amor y parentesco en la campaña de Buenos Aires en el siglo XIX," *Boletín del Instituto de Historia Argentina y Americana "Dr. Emilio Ravignani,"* 3rd ser., 30 (2007): 77.
68. Argeri, *De guerreros a delicuentes*, 231.
69. Zeballos, *Callvucurá*, 7.
70. Santiago Avedaño, *Memorias del ex-cautivo Santiago Avedaño*, ed. P. Meinrado Hux (Buenos Aires: Elefante Blanco, 1999); and Santiago Avedaño, *Usos y costumbres de los indios de la Pampa*, ed. P. Meinrado Hux (Buenos Aires: Elefante Blanco, 2000).
71. Avedaño, *Memorias*, 69, 112.
72. Avedaño, *Usos*, 45.
73. Avedaño, *Memorias*, 43.
74. Beatriz Díez, "Las Memorias de Santiago Avedaño y la trilogía de E. Zeballos." In *V Congreso Internacional de Letras*, ed. Américo Cristófalo (Buenos Aires: Facultad de Filosofía y Letras de la Universidad de Buenos Aires, 2012), 1043, *2012.cil.filo.uba.ar/sites/2012.cil.filo.uba.ar/files/0134%20DIEZ,%20BEATRIZ.pdf*.
75. For more on the trope of the white captive in fiction, see Rotker's *Captive Women* and Operé's *Indian Captivity in Spanish America*.
76. Rotker, *Captive Women*, 33.
77. Florencia Bonelli, *Indias blancas: La vuelta del ranquel* (Buenos Aires: Suma de letras, 2012).
78. Torre, *Literatura en tránsito*, 134, 148.
79. Although Mansilla claims he is acting in an official capacity, the motivations of his excursion are murkier. Silvia Fernández has demonstrated that Mansilla did not have authorization for the excursion and went to Leubucó at least partially to

avoid facing court martial for the improper execution of a deserter. Silvia Mirta Beatriz Fernández, "Mansilla y los ranqueles. ¿Porqué Lucio V. Mansilla escribió *Una excursión a los indios ranqueles?*" *Congreso Nacional de Historia Sobre la Conquista del Desierto* (Buenos Aires: Academia Nacional de la Historia, 1981), 4:370. Regardless, both in the text itself and in its reception, Mansilla's political and scientific roles cannot be separated.

80. In Antonio Espinosa, *La Conquista del Desierto. Diario del capellán de la Expedición de 1879, Monseñor Antonio Espinosa, más tarde arzobispo de Buenos Aires*, ed. Bartolomé Galindez (Buenos Aires: Freeland, 1968), 24–25.

81. In Marcela Tamagnini, *Soberanía: Territorialidad indígena. Cartas civiles II* (Temuco: Centro de Documentación Mapuche, 2003), 57. *www.mapuche.info/wps_pdf/tamagnini031104.pdf*.

82. "Guerras fronterizas," *La Prensa*, March 1, 1878.

83. Miguel Malarín to Julio A. Roca, 3 July 1879, sala VII legajo 7, Julio Argentino Roca archive, Archivo General de la Nación, Buenos Aires, Argentina.

84. Miguel Malarín to Julio A. Roca, 10 July 1879, sala VII legajo 7, Julio Argentino Roca archive, Archivo General de la Nación, Buenos Aires, Argentina.

85. Claudia Briones and José Luis Lanata, "Living on the Edge," in *Archaeological and Anthropological Perspectives on the Native Peoples of Pampa, Patagonia and Tierra del Fuego to the Nineteenth Century*, ed. Claudia Briones and José Luis Lanata (Westport: Bergin & Garvey, 2002), 9.

86. Mariano Nagy and Alexis Papazian, "El campo de concentración de Martín García: Entre el control estatal dentro de la isla y las prácticas de distribución de indígenas (1871–1886)," *Corpus: Archivo virtual de la alteridad americana* 1, no. 2 (2011): 1–22, *ppct.caicyt.gov.ar/index.php/corpus/article/view/392*.

87. Mases, *Estado y cuestión indígena*, 120.

88. Matthew Gutmann, introduction to *Changing Men and Masculinities in Latin America* (Durham: Duke University Press, 2002), 13.

89. Zeballos, *La conquista*, 280.

90. Ricardo D. Salvatore, *Wandering Paysanos: State Order and Subaltern Experience in Buenos Aires During the Rosas Era* (Durham, NC: Duke University Press, 2003), 239.

91. Mases, *Estado y cuestión indígena*, 175.

92. Mases, *Estado y cuestión indígena*, 170.

93. Nicanor Larrain, *Viajes en el "Villarino" a la costa sud de la República Argentina* (Buenos Aires: Juan A. Alsina, 1883), 61–62.

94. Same-sex relations are completely absent from contemporary texts. We are left to wonder what those relationships looked like in indigenous society before and after the Conquest of the Desert, as well as if or how creole elites reacted to them.

95. Miguel Malarín to Julio A. Roca, 25 December 1878, sala VII legajo 6, Julio Argentino Roca archive, Archivo General de la Nación.

96. The United States, Australia, and Canada—all settler colonies like Argentina—experimented with separating children from their mothers during approximately the same time period. In the United States, twenty-five federally run off-reserva-

tion boarding schools and many more on-reservation schools educated over twenty thousand indigenous children by 1899. John Bloom, *To Show What an Indian Can Do: Sports at Native American Boarding Schools* (Minneapolis: University of Minnesota Press, 2000), xii. Similarly, Canadian indigenous boarding schools began around the same time and educated a total of 150,000 children by the end of the twentieth century. Donna Feir, "The Long-Term Effects of Forcible Assimilation Policy: The Case of Indian Boarding Schools," *Canadian Journal of Economics* 49, no. 2 (2016): 436. In Australia, by 1911 the removal of indigenous children was legal in every state but one Margaret Jacobs, "Maternal Colonialism: White Women and Indigenous Child Removal in the American West and Australia, 1880–1940," *Western Historical Quarterly* 36, no. 4 (2005): 458.

97. Vicente G. Quesada, *Memoria del ministro secretario de gobierno de la Provincia de Buenos Aires presentada a las honorables Cámaras Legislativas 1877* (Buenos Aires: Imprenta y Librerias de Mayo, 1877), xxxv–xxxvi.

98. María Andrea Nicoletti, "La congregación salesiana en la Patagonia: 'Civilizar,' educar y evangelizar a los indígenas (1880–1934)," *Estudios Interdisciplinarios de América Latina y el Caribe* 15, no. 2 (2004), eial.tau.ac.il/index.php/eial/article/view/894; Yamila Liva and Teresa Laura Artieda, "Proyectos y prácticas sobre la educación de los indígenas: El caso de la misión franciscana de Laishí frente al juicio del Inspector José Elías Niklison (1901–1916)," *Historia de la educación* 15, no. 1 (2014): 68.

99. Valko, *Cazadores*, 119.

100. Emile Daireaux, *Vida y costumbres en el Plata: Tomo primero, La sociedad argentina* (Buenos Aires: Félix Lajouane, 1888), 85–86, *archive.org/details/vidaycostumbres-oodairgoog.*

101. Mases, *Estado y cuestión indígena*, 132.

102. Jorge Bustos and Leonardo Dam, "El Registro de Vecindad del Partido de Patagones (1887) y los niños indígenas como botín de guerra," *Corpus. Archivo virtual de la alteridad americana* 2, no. 1 (2012): 2, *ppct.caicyt.gov.ar/index.php/corpus/article/view/1413.*

103. Bustos and Dam, "El Registro de Vecindad," 3.

104. Mases, *Estado y cuestión indígena*, 291.

105. "Las cartas de Catriel," *El Nacional*, November 30, 1878.

106. "Las cartas de Catriel," *El Nacional*, November 30, 1878.

107. "Los indios," *La Prensa*, December 3, 1878.

108. "Esclavos en Tucumán," *La Nación*, December 25, 1888.

109. Mases, *Estado y cuestión indígena*, 110.

110. Valko, *Pedagogía*, 219.

111. "Indios en el cuartel del 8º," *El Nacional* (Buenos Aires), March 20, 1885. Given the date of this article, it is likely that the "various chieftains with their families" described by the author included Inacayal and Foyel, studied here in Chapter Three.

112. Argentina, Cámara de Diputados, *Diario de sesiones de la Cámara de Diputados: Año 1885* (Buenos Aires: Moreno y Nuñez, 1886), 2:799, 814.

113. Argentina, *Diario de sesiones* [. . .]: *Año 1885*, 2:817.

114. Argentina, Cámara de Diputados, *Diario de sesiones de la Cámara de Diputados: Año 1888* (Buenos Aires: Sudamericana, 1889), 1:101.

115. Mases, *Estado y cuestión indígena*, 126.

116. Daireaux, *Vida y costumbres*, 52; Argentina, *Diarios de Sesiones* [. . .]: *Año 1888*, 1:101.

117. Indeed, the sixteenth-century priest Bartolomé de las Casas decried the frequent rape of indigenous women across the Spanish Empire. Bartolomé de las Casas, *Brevísima relación de la destruición de las Indias*, ed. André Saint-Lu (Madrid: Cátedra, 1989), 120.

CHAPTER 2

1. For more on this paradox, see Rebecca Earle, *The Return of the Native: Indians and Myth-Making in Spanish America, 1810–1930* (Durham, NC: Duke University Press, 2007), and Larson, *Our Indigenous Ancestors.*

2. Francisco P. Moreno to Pedro Pico, 14 September 1875, in *Reminiscencias de Francisco P. Moreno: Versión propia*, ed. Eduardo V. Moreno (Buenos Aires: Editorial Universitaria de Buenos Aires, 1979), 20.

3. John Lubbock established these parallels in *The Origin of Civilisation.* In the River Plate, the perceived equivalence of prehistoric and present-day peoples can be clearly seen in Francisco P. Moreno's assertion that, in Argentina, the explorer could easily "travel from the refinement of civilization and science, to fossil times. In the course of two months the traveler can palpably traverse 200,000 years and can see his grandfather armed sometimes with a sharp stone, fighting beasts for his food, and other times, combating them with the steel weapons that his grandson, carried by the irresistible force of progress, has succeeded in forging, transforming, with the evolution of his intelligence, the flint arrowhead or knife." Moreno, *Viaje a la Patagonia austral*, 227–28.

4. Moreno, *Viaje a la Patagonia austral*, 92–93. Similarly, in 1875 Moreno wrote to his father, "Although I think I will not be able to complete the number of skulls I desired, I am sure that tomorrow I will have 70." As if reading from a shopping list, he named the skulls he had collected and those he hoped to soon acquire: he already owned the skulls of Cipriano Catriel and his wife, "and now, it seems that the younger brother Marcelino will not live much longer." Francisco P. Moreno to Francisco Moreno, 5 April 1875, in Moreno, *Reminiscencias*, 53.

5. Peñaloza, "On Skulls," 467.

6. Peter Pesic, "Desire, Science, and Polity: Francis Bacon's Account of Eros," *Interpretation: A Journal of Political Philosophy* 26, no. 3 (Spring 1999): 342. In the present, sociologists of scientific knowledge and scholars of science, technology, and society have shown that the construction and distribution of scientific knowledge is fundamentally social, rooted in personal relations, values, and local contexts. Anthropology and ethnography are particularly tense fields, as the nature of fieldwork requires the foregrounding of personal experience even while eschewing it

in the name of objectivity. While theorists have begun to investigate desire, intimacy, and even sexual relationships within present-day ethnography, these ideas have appeared only recently in histories of the field.

7. Jorge Carman, introduction to *Obras*, by Ramón Lista, ed. Jorge Carman (Buenos Aires: Confluencia, 1998), 1:15.

8. Ramón Lista, *La Gobernación de Santa Cruz*, 2:189.

9. Lista, *Viaje al país de los onas*, 2:18.

10. Mariela Eva Rodriguez's investigation of censuses in the Dirección Nacional de Tierras y Colonias in the Consejo Agrario Provincial de Santa Cruz indicate that Ramona was fifty-four in 1943 and also in 1944, giving a birthdate of 1890 or 1891. Mariela Eva Rodríguez, email message to author, October 13, 2017. See also Rodríguez, "De la 'extinción' a la autoafirmación," 148–60. Other historians agree with Rodriguez's reading, except Osvaldo Mondelo who raises doubts about Ramona's birthdate and whether Lista ever met her in his book *Tehuelches: Danza con fotos.* Referring to baptismal records from Santa Cruz, Mondelo argues that Ramona was born in May 1893, six months after Lista left Santa Cruz. Osvaldo L. Mondelo, *Tehuelches: Danza con fotos* (El Calafate: el autor, 2012), 134.

11. D. Osvaldo Topcic', *Historia de la provincia de Santa Cruz: Crónicas y testimonios* (Córdoba: Centro de Estudios Históricos "Prof. Carlos S. A. Segreti," 2006), 301.

12. Eduardo Pogoriles, "Ramón Lista, el hombre enamorado de los tehuelches," *Clarín*, May 12, 2001, *www.clarin.com/sociedad/ramon-lista-hombre-enamorado-tehuelches_0_S13GozdlAYg.html.*

13. Roberto Payró, *La Australia argentina*, 5th ed. (Buenos Aires: La Nación, 1898), 93.

14. Juan Hilarión Lenzi, "Ramón Lista, gobernador de Santa Cruz," *Argentina austral* 302 (December 1956).

15. Neither the biography nor the explanatory passages in *Obras*, Lista's collected works compiled by his great-grandson, Jorge Carman, make any reference to the scandal. That period of his life is simply passed over. One of the few texts to explore the relationship between Lista and Coile is María Rosa Lojo's *Historias ocultas en la Recoleta*, a collection of fictionalized histories of the inhabitants of La Recoleta, Argentina's most famous cemetery. The two chapters relating to Lista look at Andrade's suicide from the perspective of her mother and that of Lista himself. While interesting, the text is not intended to be historically accurate, as Lojo declares her dedication "to the liberty and aesthetic autonomy of fiction." María Rosa Lojo, *Historias ocultas en la Recoleta* (Buenos Aires: Suma de Letras, 2002), 27.

16. Charles Darwin, *The Voyage of the Beagle* (Auckland: Floating Press, 2008), 334–35 and 348–49.

17. For more on how this discourse functioned, see Quijada, "Ancestros, ciudadanos, piezas de museo" and Navarro Floria, Salgado and Azar, "La invención de los ancestros."

18. Ramón Lista, *El Territorio de las Misiones*, 1:338. Italics in the original.

19. Johannes Fabian, *Time and the Other: How Anthropology Makes its Object* (New York: Columbia University Press, 1983), 31.

20. Lista, *Mis exploraciones y descubrimientos en la Patagonia*, 1:195.
21. Lista, *Viaje al país de los tehuelches*, 1:45, 90.
22. Lista, *Viaje al país de los tehuelches*, 1:58.
23. Lista, *Viaje al país de los tehuelches*, 1:68.
24. Lista, "El país de las Manzanas," 1:263.
25. Lista, *Los indios tehuelches*, 2:127.
26. Lista, *Mis exploraciones*, 1:178; Lista, *Los indios tehuelches*, 2:137–140.
27. Musters explains that Orkeke's niece, "a remarkably pretty little girl of about thirteen years of age . . . instructed me in the Indian names of the various objects about the place." Musters, *At Home with the Patagonians*, 62. Mansilla calls Carmen his "instructress" in Araucanian. Mansilla, *A Visit*, 224.
28. Knowledge of his affair also lends a particular irony to certain passages. For example, he argues that adultery is almost unknown in Tehuelche society, except among those suffering from "a special pathological state or a certain perversion of moral sense to which the Christians that live in the surrounding areas or in the *toldos* themselves have recently contributed—Christians who are the cream of the crop in terms of corruption and robbery." Lista, *Los indios tehuelches*, 2:155.
29. Payró, *La Australia argentina*, 96.
30. For more on this dynamic in the work of Lista, Payró, and Clemente Onelli, see Ashley Elizabeth Kerr, "Progress at What Price?: Defenses of Indigenous Peoples in Argentine Writing about Patagonia (1894–1904)," *Decimonónica* 16, no. 2 (Summer 2019): 17–33.
31. Lista, *Mis exploraciones*, 1:195.
32. See, for example, Paul Broca's summary of the research in Paul Broca, *On the Phenomena of Hybridity in the Genus Homo*, ed. and trans. C. Carter Blake (London: Longman, Green, Longman, and Roberts, 1864), 28–29. *HathiTrust, hdl.handle.net/2027/nyp.33433061699843*.
33. Matthew Cobb, "Heredity before Genetics: A History," *Nature Reviews: Genetics*, 7 (December 2006): 954.
34. Ernst Haeckel, *The History of Creation, or, The Development of the Earth and Its Inhabitants by the Action of Natural Causes*, trans. E. Ray Lankester, vol. 1 of 2 (London: Henry S. King & Co., 1876), 211.
35. Eva Gillies, introduction to *A Visit to the Ranquel Indians*, ed. and trans. Eva Gilles (Lincoln: University of Nebraska Press, 1997), xix.
36. Fernández, "Mansilla y los ranqueles," 370.
37. Carlos Alonso, "Oedipus in the Pampas: Lucio Mansilla's *Una Excursión a Los Indios Ranqueles*," *Revista de Estudios Hispánicos* 24, no. 2 (May 1990): 44; J. Andrew Brown, *Test Tube Envy: Science and Power in Argentine Narrative* (Lewisburg: Bucknell University Press, 2005), 56–58; and Homero M. Guglielmini, *Mansilla* (Buenos Aires: Ediciones Culturales Argentinas, 1961), 83.
38. Gillies, introduction to *A Visit*, xxxv–xxxvi.
39. Andrew Brown, *Test Tube Envy*, 58.

40. Anne McClintock, *Imperial Leather: Race, Gender, and Sexuality in the Colonial Contest* (New York: Routledge, 1995), 22.

41. Samuel Monder, "La ley del deseo: Acerca de 'Una excursión a los indios ranqueles,' de Lucio V. Mansilla," *Iberoamericana (nueva Época)* 8, no. 32 (December 2008): 63. Mansilla writes of the Ranqueles: "There is no Indian who does not possess one or several secrets, perhaps without importance, but that he will not reveal except when it is in his interest to do so." *A Visit*, 111. He also notes that women keep things to themselves and prefer to sigh mysteriously to speaking plainly (333).

42. McClintock, *Imperial Leather*, 23. The application of McClintock's ideas to readings of Mansilla is especially interesting given that Mansilla owned a copy of *King Solomon's Mines,* the Rider Haggard book McClintock uses to illustrate many of her ideas. For more on Mansilla's library, see María Montero, "Mansilla y sus bibliotecas," *Boletín de La Academia Argentina de Letras* 56 (1991): 103–28.

43. Rousseau was one of Mansilla's preferred readings: he describes his first experience with the *Social Contract* in several of his *causeries* and frequently mentions Rousseau in *Una excursión.*

44. Lista, *Viaje al país de los tehuelches*, 1:73; Lista, *Mis exploraciones*, 1:125.

45. Orión refers to Héctor Florencio Varela, the director of *La Tribuna*, the newspaper in which Mansilla published his letters. Carlos Keen was an Argentine lawyer and journalist. Calfucurá was a Mapuche chieftain, while Mariano Rosas was the chieftain of the Ranqueles, the group Mansilla was studying.

46. Cristina Iglesia, "Mejor se duerme en la pampa. Deseo y naturaleza en *Una excursión a los indios ranqueles* de Lucio V. Mansilla," *Revista Iberoamericana* 63, no. 178-179 (June 1997): 188.

47. Santiago Arcos, "Cuestión de indios" in *Cuestión de indios*, ed. Juan Bautista Sasiaiñ (Buenos Aires: Policía Federal Argentina, 1979), 7–30; and Santiago Arcos, *La Plata: Etude historiqué* (Paris: Michel Levy Freres, 1865).

48. George W. Stocking, Jr., *Race, Culture, and Evolution. Essays in the History of Anthropology* (New York: Free Press, 1968), 39.

49. Montero, "Mansilla y sus bibliotecas," 126.

50. María Rosa Lojo, "El indio como 'prójimo,' la mujer como el 'otro' en *Una excursión a los indios ranqueles* de Lucio V. Mansilla," *Alba de América* 26-27 (July 1996): 134.

51. Lojo, "El indio como 'prójimo,'" 132.

52. Javier Gasparri, "Mansilla: La promiscuidad de los cuerpos," in *II Coloquio internacional: Saberes contemporáneos desde la diversidad sexual: Teoría, crítica, praxis* (Rosario: Universidad Nacional de Rosario, 2013), 100; and Monder, "La ley del deseo," 72. Gasparri even opens his article with an anecdote attributed to Victoria Ocampo implying that Mansilla once tried homosexual sex (102).

53. Broca, *On the Phenomena of Hybridity*, 15.

54. Broca, *On the Phenomena of Hybridity*, x.

55. Argentina, Cámara de Diputados, *Diario de sesiones de la Cámara de Diputados: Año 1885* (Buenos Aires: Moreno y Nuñez, 1886), 1:503.

56. Argentina, Cámara de Diputados, *Diario de sesiones de la Cámara de Diputados: Año 1885*, 1:506.
57. Argentina, Cámara de Diputados, *Diario de sesiones de la Cámara de Diputados: Año 1885*, 1:504.
58. Mases, *Estado y cuestión indígena*, 204.
59. Jager, *Malinche*, 122. In the case of Argentina, see Roulet, "Mujeres, rehenes y secretarios," 303–37.
60. Jager, *Malinche*, 5.
61. Young, *Colonial Desire*, 25.
62. Maureen Spillane McKenna, "El debate del ser nacional argentino: La frontera en las obras de Lucio Mansilla y César Aira," *Hispanófila* 142 (September 2004): 94.
63. Philip J. Deloria, *Playing Indian* (New Haven: Yale University Press, 1998), 185.
64. Deloria, *Playing Indian*, 78. Mansilla's eighteen days of playing Indian are dwarfed by the playacting of the US ethnographer Lewis Henry Morgan in the 1840s. Along with several friends, Morgan founded a cultured society that quickly transformed into the New Confederacy of the Iroquois, a fraternal organization dedicated to gathering information about the Iroquois. More unusual, however, was the organization's dedication to playing Indian, as members wore Iroquois-style clothing, learned the language, and were inducted into the group through an "Inindianation" ceremony in which they were transformed "from Yankees—the actual beneficiaries of American Indian policy—to aboriginal American Selves," adopting Iroquois names to reflect their new spirit.
65. Deloria, *Playing Indian*, 184.
66. Nouzeilles, "El retorno de lo primitivo," 94.
67. Gary Alan Fine, "Ten Lies of Ethnography: Moral Dilemmas of Field Research," *Journal of Contemporary Ethnography* 22 (April 1993): 283. The most complete approach to these issues can be found in Don Kulick and Margaret Willson, eds., *Taboo: Sex, Identity, and Erotic Subjectivity in Anthropological Fieldwork* (London: Routledge, 1995).
68. Robert Bieder, *Science Encounters the Indian, 1820-1880: The Early Years of American Ethnology* (Norman: University of Oklahoma Press, 1986), 162.
69. Deborah Reed-Danahay, "Autobiography, Intimacy and Ethnography," in *Handbook of Ethnography*, ed. Paul Atkinson et al. (London: Sage Publications, 2001), 415.

CHAPTER 3

1. Moreno to Marcelino Vargas, 2 October 1886, in *Reminiscencias*, 207. As he writes in the letter "After I have them gathered here, where they will have good rooms, horses, etc., I will try to find the way to alleviate the fate of the good Indians who are still in el Tigre." In a letter the same day to the Gefe [*sic*] de Talleres del Tigre where the Tehuelches were being held, Moreno specifies that the fifteen are "Inacayal a brother his wife and 3 or 4 children, Foyel, his brother, his wife and children and the interpreter that accompanies them—in all 15 people between the young and old" (Moreno to Antonio Muratorio, 2 October 1886, in *Reminiscencias*,

208). These statements, along with the photographic record, suggest that the family was in fact split up.

2. Inacayal and Foyel's experience is somewhat of an extreme example, but not unique. Orkeke, Shaihueque, Pincén, and other caciques were also brought to Buenos Aires. During these stays, they met with political elites (including the president) and experienced events like Carnival and ice skating, delighting local audiences with the seemingly incongruous marriage of civilization and barbarism. Although their encounters with science were not as intense as those of Foyel and Inacayal, many of these chieftains interviewed with Moreno or Lista (at the government's or their request) and were also photographed by leading photographers of the time. These images subsequently formed part of the anthropological-historical record, serving as evidence of the characteristics of Argentine indigenous peoples.

3. Quijada, "Ancestros, ciudadanos, piezas de museo," 10.

4. Carolyne Ryan Larson is one of the few to highlight the women's experience. Larson, *Our Indigenous Ancestors*, 43–44. Ana Butto has looked at gendered aspects of photographs of Tehuelche and Mapuche Indians in the late nineteenth century more broadly. Ana Butto, "Visualización de los roles de género en las fotografías etnográficas de mapuches y tehuelches (siglos XIX-XX)," *RiHumSo* 7, no. 13 (November 2018): 21–46.

5. While I refer to them as Tehuelches, following Moreno, several of the members of the group were likely from other tribes. Milcíades Alejo Vignati, a curator at the Museo de La Plata in the 1930s and 1940s, attempted to identify their origins in "Iconografía aborigen I: Los caciques Sayeweke, Inakayal y Foyel y sus allegados," *Revista del Museo de La Plata (nueva Serie)* 2, no. Antropología, no. 10 (1942): 13–48.

6. We do not know exactly where these photographs were taken. The most likely location is the museum itself, given the timing and Vignati's claim that some of the chairs seen in the images were still in the museum when he was writing in the 1940s. Vignati, "Iconografía aborigen I," 16.

7. Elizabeth Edwards, "The Image as Anthropological Document. Photographic 'Types': The Pursuit of Method," *Visual Anthropology* 3 (1990): 235.

8. Marta Penhos, "Frente y perfil: Una indagación acerca de la fotografía en las prácticas antropológicas y criminológicas en Argentina a fines del siglo XIX y principios del XX," in *Arte y antropología en la Argentina*, eds. Marta Penhos and Marina Baron Supervielle (Buenos Aires: Fundación Espigas, 2005), 22.

9. Farro, *La formación*, 76.

10. Farro, "Imágenes de cráneos, retratos antropológicos y tipologías raciales: Los usos de las primeras colecciones de fotografías del Museo de La Plata a fines del siglo XIX," in *Los secretos de Barba Azul: Fantasías y realidades de los archivos del Museo de La Plata*, eds. Tatiana Kelly and Irina Podgorny (Rosario: Prohistoria, 2012), 78.

11. Elizabeth Edwards, "Material Beings: Objecthood and Ethnographic Photographs," *Visual Studies* 17, no. 1 (2002): 68.

12. Some of the images were captioned during Moreno's time, others by Robert Lehmann-Nitsche between 1898 and 1930. Yet others were identified by Milcíades

Vignati, curator at the museum in the 1940s. Máximo Farro, "Imágenes de cráneos," 87. Martha Bechis argues that it was not uncommon for even the most famous indigenous individuals in nineteenth-century photographs to be mislabeled; it is thus entirely probable that several of the Tehuelches were erroneously named or destined to remain anonymous, stripped of their personhood. Bechis, "Rostros aborígenes de las pampas argentinas, siglos XVIII–XIX," *Boletín* TEFROS 2, no. 2 (Winter 2004): n.p., *www.hum.unrc.edu.ar/ojs/index.php/tefros/article/view/78/61*. Finally, Vignati cropped or otherwise altered many of the images in order to make them more closely resemble anthropometrical or criminological photos. When Vignati was let go from the museum in 1955, he took many of the printed albumen images with him, leaving only twelve. Farro, "Imágenes de cráneos," 80–84. The images that Vignati took are now in the "Colección Vignati" in the Biblioteca Popular "Agustín Alvarez" in Trelew.

13. These include six negatives on glass slides in the Archivo Fotográfico General (AFO), twelve albumen prints of approximately 15 × 20 cm on 21 × 42 cm cardboard sheets in the Boote collection (BOO), and seventy-five negatives in the Colecciones de la Sección Arqueologíca (ARQ). The catalog of images can be found on the CD which accompanies Irina Podgorny and Tatiana Kelly, eds., *Los secretos de Barba Azul: Fantasías y realidades de los archivos del Museo de La Plata* (Rosario: Prohistoria, 2012) or on the Museo de La Plata's website: *www.museo.fcnym.unlp.edu.ar/archivo_historico_colecciones*.

14. While the two groups overlapped slightly in Buenos Aires, Moreno claims that Shaihueque did not know that Inacayal and Foyel were in the city. Francisco P. Moreno, "Inacayal y Foyel," *El Diario* (Buenos Aires), March 17, 1885. The date of the invoice is months after Shaihueque's departure from Buenos Aires and thus it is unlikely to include the images of him and his tribe. Alejandro Martínez, "Imágenes fotográficas sobre pueblos indígenas: Un enfoque antropológico" (PhD diss., Universidad Nacional de La Plata, 2010), 16. Furthermore, the images I have identified as being of Inacayal, Foyel, and their group number almost exactly forty-seven, the number of images indicated on the invoice. Thus, we can conclude that the photos that can be attributed to Boote are the images with the rougher floor, lower transition from stone to plaster, and curved-back chair, likely taken in Retiro in or around July 1885. The others probably were taken in the same space by an unknown photographer sometime between February 22 and April 1, 1885.

15. Ana Butto, "Con el foco en el otro: Las representaciones visuales acerca del indio y el territorio en los expedicionarios de la Conquista Del Desierto en las campañas de 1879 y 1883," in *Entre pasados y presentes* III*: Estudios contemporáneos en ciencias antropológicas*, ed. Kuperszmit et al. (Buenos Aires: Mnemosyne, 2012), 116.

16. Butto, "Con el foco," 117.

17. Carlos Masotta, *Indios en las primeras postales fotográficas argentinas del s. XX* (Buenos Aires: La Marca Editora, 2007), 11–12.

18. Penhos, "Frente y perfil," 43; and Farro, "Imágenes de cráneos," 82. Numerous scholars argue that this style indicates that Moreno was attempting to repro-

duce Bertillon-style criminological photography, thus criminalizing the Indian and supporting state efforts to control him. Penhos, "Frente y perfil," 35; Ana R. Butto and Dánae Fiore, "Violencia fotografiada y fotografías violentas: Acciones agresivas y coercitivas en las fotografías etnográficas de pueblos originarios fueguino y patagónicos," *Nuevo Mundo Mundos Nuevos* (2014): 12; and Sergio Caggiano, "La visión de la 'raza': Apuntes para un estudio de la fotografía de tipos raciales en Argentina," *Revista del Museo de Antropología* 6 (2013): 111. Nonetheless, Máximo Farro has convincingly demonstrated that the mug shot style of the images was a result of Vignati's interventions in the 1940s, not Moreno's instructions. It is also unlikely that Moreno would have known Bertillon's work in the first half of 1885. Farro, "Imágenes de cráneos," 83–84. The November 1885 Congress of Criminal Anthropology in Rome is likely the first time that South Americans became aware of Bertillon's ideas. Mercedes García Ferrari and Diego Galeano, "Police, anthropometry, and fingerprinting: Transnational History of Identification Systems from Rio de la Plata to Brazil," *História, Ciências, Saúde–Manguinhos* 23 (Dec. 2016): 6. *www.scielo.br/pdf/hcsm/v23s1/en_0104-5970-hcsm-23-s1-0171.pdf.*

19. In this regard, the images contrast greatly with the elaborate adornments seen in contemporary field photography from Chile, for example. See Margarita Alvarado Pérez, Pedro Mege, and Christian Báez, eds., *Mapuche fotografías siglos XIX y XX: Construcción y montaje de un imaginario* (Santiago: Pehuén, 2001).

20. Farro, "Imágenes de cráneos," 80.

21. Andermann, *The Optic of the State*, 16.

22. Christopher Matthews, "History to Prehistory: An Archaeology of Being Indian in New Orleans, *Archaeologies: The Journal of the World Archeological Congress* 3, no. 3 (2007): 288.

23. Henrietta Riegel, "Into the Heart of Irony: Ethnographic Exhibitions and the Politics of Difference," in *Theorizing Museums*, ed. Sharon Macdonald and Gordon Fyfe (Oxford: Blackwell, 1996), 86.

24. Edwards, "The Image as Anthropological Document," 241. In Argentina, subjects were often labeled "indios" (Indians) or "indios típicos" (typical Indians). Mariana Giordano, *Indígenas en la Argentina: Fotografías 1860–1970* (Buenos Aires: El Artenauta, 2012), 17.

25. The white bars around her face reflect Milcíades Alejo Vignati's cropping in the early 1940s. Farro, "Imágenes de cráneos," 86.

26. Margarita Alvarado Pérez, "Pose y montaje en la fotografía *mapuche*: Retrato fotográfico, representación e identidad," in *Mapuche fotografías siglos XIX y XX: Construcción y montaje de un imaginario*, ed. Margarita Alvarado Pérez, Pedre Mege, and Christian Báez (Santiago: Pehuén, 2001), 15; and Giordano, *Indígenas en la Argentina*, 34.

27. Moreno, "Inacayal y Foyel."

28. Moreno, "Inacayal y Foyel." Pincén and Namuncurá were caciques who actively resisted Argentine control. Orkeke was another Patagonian chieftain who was brought to Buenos Aires. The *porteño* newspapers delighted in reporting his

visits to banquets, the opera, and the theater (Valko, *Cazadores de poder*, 149–63). In the end, Orkeke and many members of his tribe died in Buenos Aires before they could return to their homeland (Valko, *Cazadores de poder*, 179).

29. Efram Sera-Shriar, "Anthropometric Portraiture and Victorian Anthropology: Situating Francis Galton's Photographic Work in the Late 1870s," *History of Science* 53, no. 2 (2015): 161.

30. Herman ten Kate, "Matériaux pour servir à l'anthropologie des Indiens de la République Argentine," *Revista del Museo de La Plata* 12 (1906): 51. Moreno himself analyzes Tehuelche men and women's measurements separately in *Viaje a la Patagonia austral*, 377–78.

31. Gastón Gordillo and Silvia Hirsch, "Indigenous Struggles and Contested Identities in Argentina: Histories of Invisibilization and Reemergence," *The Journal of Latin American Anthropology* 8, no. 3 (2003): 11; Larson, *Our Indigenous Ancestors*, 20.

32. Butto, "Visualización" 26.

33. Mansilla, *A Visit*, 326.

34. Butto, "Visualización," 34-35.

35. Philippa Levine, "States of Undress: Nakedness and the Colonial Imagination," *Victorian Studies* 50, no. 2 (Winter 2008): 189.

36. Levine, "States of Undress," 207.

37. Sander L. Gilman, "Black Bodies, White Bodies: Toward an Iconography of Female Sexuality in Late Nineteenth-Century Art, Medicine, and Literature," *Critical Inquiry* 12, no. 1 (Autumn 1985): 216. See also Sadiah Qureshi, "Displaying Sara Baartman, the 'Hottentot Venus,'" *History of Science* xlii (2004): 233–57.

38. Somerville, "Scientific Racism," 251–52.

39. Butto, "Visualización," 33.

40. Levine, "States of Undress," 199.

41. Moreno, *Viaje a la Patagonia austral*, 233–34.

42. Moreno, *Viaje a la Patagonia austral*, 13, 215.

43. Lucio Mansilla describes this custom among the Ranqueles in *A Visit*, 224. According to Argeri, this practice also extended to other tribes of Norpatagonia. *De guerreros a delincuentes*, 231.

44. Lista, *Viaje al país de los tehuelches*, 42; Lista, *Mis exploraciones*, 180.

45. Moreno, *Viaje a la Patagonia austral*, 367.

46. Moreno, *Reminiscencias*, 118.

47. Mansilla, *A Visit*, 301.

48. Moreno, *Reminiscencias*, 172.

49. Moreno, "Inacayal y Foyel."

50. T. H. Holdich, "Obituary: Dr. Francisco P. Moreno, Gold Medalist of the Society and Honorary Corresponding Member," *Geographical Journal* 55, no. 2 (February 1920): 157.

51. Butto and Fiore, "Violencia fotografiada," 12.

52. Edwards, "The Image as Anthropological Document," 237.

53. Dánae Fiore and María Lydia Varela, *Memorias de papel: Una arqueología visual de las fotografías de pueblos originarios fueguinos* (Buenos Aires: Dunken, 2009), 25.
54. A decade later, in 1896, a group of Argentine Indians refused to be photographed in the Museo de La Plata not due to fear of the camera, but because two of them worked as policemen and were acutely aware of Alphonse Bertillon's criminological uses of the technology. Ten Kate, "Matériaux," 52. Argentina was the first country outside of France to establish an official anthropometric service following the precepts of *bertillonage*. García Ferrari and Galeano, "Police, Anthropometry, and Fingerprinting," 7.
55. Ana Butto, "Artefactos autóctonos y foráneos en las fotografías de indígenas y criollos militares durante la 'Conquista Del Desierto' (Norpatagonia, Siglo XIX)," in *Tendencias teórico-metodológicas y casos de estudio en la arqueología de la Patagonia*, ed. Atilio Zangrando et al (San Rafael: Museo de Historia Natural de San Rafael, 2013), 57.
56. Sera-Shriar, "Anthropometric Portraiture and Victorian Anthropology," 161.
57. Anne Matthews, "Modern Anthropology and the Problem of the Racial Type: The Photographs of Franz Boas," *Visual Communication* 12, no. 1 (February 2013): 128.
58. The nude images of men are not as strongly segregated by class. The cacique Foyel is photographed shirtless, and Inacayal's cousin Sayeñamku is photographed both shirtless and fully nude. The fact that elite male Tehuelches were forced to be photographed nude is likely because these images were taken within the museum where there was a small pool of subjects to draw from, and all of them were closely related to Inacayal or Foyel because Moreno could not (or would not) negotiate to bring the others to the museum.
59. Robert Lehmann-Nitsche, "Etudes anthropologiques sur les indiens ona (groupe tshon) de la Terre de Feu," *Revista del Museo de La Plata* 23 (second part) (1916): 174.
60. Michael Kraus, "Of Photography and Men: Encounters with Historical Portrait and Type Photographs," in *Photography in Latin America: Images and Identities across Time and Space*, ed. Gisela Cánepa Koch and Ingrid Kummels (Bielefeld: Transcript, 2016), 41.
61. Moreno, *Reminiscencias*, 180.
62. While the predominant stereotype is of indigenous peoples fleeing the camera lens in fear, scholars such as Sandweiss and Cohan Scherer analyze several examples of indigenous peoples in the United States seeking out portrait photography for a variety of purposes. Martha A. Sandweiss, *Print the Legend: Photography and the American West* (New Haven: Yale University Press, 2002); Joanna Cohan Scherer, "The Public Faces of Sarah Winnemucca," *Cultural Anthropology* 3, no. 2 (May 1988): 178–204. Similarly, K'iches in Guatemala patronized Tomás Zanotti's photography studio at least as early as 1898. Greg Grandin, "Can the Subaltern Be Seen? Photography and the Affects of Nationalism," *Hispanic American Historical*

Review 84, no. 1 (2004), 83. Some groups were quite savvy, "participating as full partners in the collaborative process of making a picture, and in understanding how photographic portraits could serve personal needs." Sandweiss, *Print the Legend*, 210. While there is no direct evidence of these interventions on the part of the Tehuelches, to assume that, unlike white Argentines, the Tehuelches had no desire to see themselves or their families in print once again postulates them as primitive, uncultured, and inferior.

63. Mansilla, *A Visit*, 186–87; Moreno, *Viaje a la Patagonia austral*, 232.

64. Mabel Fernández, "La representación de las mujeres aborígenes en la iconografía patagónica," *Aljaba* 19 (2015): n.p.

65. Moreno, *Viaje a la Patagonia austral*, 11, 446.

66. Musters, *At Home with the Patagonians*, 196.

67. Lista, *Viaje al país de los tehuelches*, 115.

68. Mases, *Estado y cuestión indígena*, 110.

69. Lehmann-Nitsche, "Etudes anthropologiques sur les indiens ona," 175.

70. See, for example, Lista, "Mis exploraciones," 182.

71. Moreno, *Viaje a la Patagonia austral*, 13.

72. Moreno, "Inacayal y Foyel."

73. Moreno, *Viaje a la Patagonia austral*, 234.

74. Moreno to Ameghino, 2 August 1887, in Florentino Ameghino, *Obras completas y correspondencia científica de Florentino Ameghino*, ed. Alfredo J. Torcelli (La Plata: Taller de Impresiones Oficiales, 1935), 20:426.

75. Ameghino to Moreno, 7 August 1887, in Ameghino, *Obras completas y correspondencia científica*, 20:427.

76. ten Kate, "Matériaux," 40.

77. Palermo, "El revés de la trama," 73-80.

78. There is some confusion about who died and when. The *La Capital* article lists Inacayal, Margarita, and a seven-year-old girl, likely the daughter of Inacayal posed standing next to a chair in the Boote photos. Inacayal and Margarita's remains were integrated into the museum collections; the seven-year-old has gone unnamed and undiscovered. The remains of Inacayal's wife and Tafá also joined the museum galleries, indicating that they, too, died during their confinement in the museum. The date of Inacayal's death is given by *La Capital* as Sept. 26, 1887, while posterior accounts by the museum say that he died on September 24, 1888. Vignati, "Iconografía aborigen I," 25.

79. "Denuncia gravísima," *La Capital* (La Plata), September 27, 1887.

80. Francisco P. Moreno, "Rectificación," *El Diario* (Buenos Aires), September 28, 1887.

81. Moreno, "Rectificación."

82. Lista, *Los indios tehuelches*, 110.

83. María Luz Endere, "Cacique Inakayal: La primera restitución de restos humanos ordenada por ley," *Corpus: Archivo virtual de la alteridad americana* 1, no. 1 (Primer semestre 2011): 2.

84. ten Kate, “Matériaux,” 51.
85. Edwards, “The Image as Anthropological Document,” 241.
86. Lucio V. Mansilla, *Una excursión a los indios ranqueles*, ed. J. Bouchet (Buenos Aires: Juan A. Alsina, 1890), n.p.; Gillies, ed., *A Visit*, n.p. Milcíades Alejo Vignati noted Bouchet’s use of the image in “Falacias iconográficas,” *Relaciones de la Sociedad Argentina de Antropología* 4 (1944): 261–62.
87. Jane Hubert and Cressida Fforde, “Introduction: The Reburial Issue in the Twenty-First Century,” in *The Dead and Their Possessions: Repatriation in Principle, Policy, and Practice*, ed. Cressida Fforde, Jane Hubert, and Paul Turnbull (London: Routledge, 2002), 12.
88. Silvia Ametrano, “Los procesos de restitución en el Museo de La Plata,” *Revista argentina de antropología biológica* 17, no. 2 (Dec. 2015): n.p.
89. Máximo Farro, “Natural History Museums in Argentina, 1862–1906,” *Museum History Journal* 9, no. 1 (January 2016): 126.
90. Farro, “Natural History Museums,” 129.

CHAPTER 4

1. Eduardo Ladislao Holmberg, “La noche clásica de Walpurgis.” *Anales de la Sociedad Científica Argentina* 22 (1886): 262.
2. Holmberg, “La noche clásica,” 264-67.
3. Ramón Lista’s “scientific fantasy,” *Quirón*, has unfortunately been lost, leaving only the fragment he included in *Los indios tehuelches*. In this passage, he recounts the arrival of Ferdinand Magellan to the Bay of San Julián (Santa Cruz Province, Argentina) from the perspective of the Patagonians. The dialogue captures their surprise:

> “Did you see that, that flew above the water like a bird?”
> “Yes,” replied *Yaten*, “it is the ‘thing’ in which the snow-faced men move about.”
> “Where will they go, *Yaten*? . . . Who knows if they might not return in the dark and try to surprise us, snatch our women, and then kill us so that the caracaras (*caranchos*) pull out our eyes and the pumas eat our hearts . . . ?”

It also projects the same view of conquest that Lista elaborates in other parts of *Los indios tehuelches*: cruelty, violence, and rape, not at the hands of the Indians but by so-called civilization. Without the rest of the text it is difficult to make definitive statements about *Quirón*, but from this small piece it appears that Lista meant it to be the fictional counterpart to the theses he puts forth in *Los indios tehuelches*. Lista, *Los indios tehuelches*, 140.

4. Fernando Unzueta, “Scenes of Reading: Imagining Nations / Romancing History in Spanish America,” in *Beyond Imagined Communities: Reading and Writing the Nation in Nineteenth-Century Latin America*, ed. Sara Castro-Klarén and John Charles Chasteen (Washington, DC: Woodrow Wilson Center Press, 2003), 122.

5. Doris Sommer, *Foundational Fictions: The National Romances of Latin America* (Berkeley: University of California Press, 1991), 7.
6. Gillian Beer, *Open Fields: Science in Cultural Encounter* (Oxford: Oxford University Press, 1996), 173.
7. Daniel Balderston, "The Indianist Novels of Estanislao S. Zeballos," *Revista Canadiense de Estudios Hispánicos* 15, no. 2 (Winter 1991): 326. *www.jstor.org/stable/27762828.*
8. Zeballos, *Painé,* 169.
9. Díez, "Las *Memorias* de Santiago Avedaño," 1042.
10. Rotker, *Captive Women*, 58–59.
11. Manuel Baigorria, *Memorias*, ed. P. Meinrado Hux (Buenos Aires: Elefante Blanco, 2006), 251–55.
12. Bjerg, "Vínculos mestizos," 77.
13. Operé, *Indian Captivity in Spanish America*, 215.
14. Balderston, "The Indianist Novels," 326.
15. Eduardo Ladislao Holmberg, *Lin-Calél* (Buenos Aires: Talleres Gráficos de L.J. Rosso, 1910), 309.
16. Azar, Nacach, and Navarro Floria, "Antropología, genocidio y olvido," 82.
17. Marcelo Montserrat, "The Evolutionist Mentality in Argentina: An Ideology of Progress," in *The Reception of Darwinism in the Iberian World: Spain, Spanish America and Brazil*, ed. Thomas F. Glick, Miguel Angel Puig-Samper, and Rosaura Ruiz (Dordrecht: Kluwer Academic Publishers, 2001), 2.
18. Novoa and Levine, *From Man to Ape*, 20.
19. Novoa and Levine, *From Man to Ape*, 84.
20. Gioconda Marún, introduction to *Olimpio Pitango de Monalia*, by Eduardo L. Holmberg, ed. Giaconda Marún (Buenos Aires: Ediciones Solar, 1994), 9.
21. Adriana Novoa and Alex Levine, *¡Darwinistas!: The Construction of Evolutionary Thought in Nineteenth-Century Argentina* (Leiden: Brill, 2012), 29.
22. José Manuel Estrada, *El génesis de nuestra raza: Refutación de una lección del Dr. D. Gustavo Minelli sobre la misma materia* (Buenos Aires: La Bolsa, 1862), 32.
23. Estrada, *El génesis,* 35, 49.
24. Thomas F. Glick, "The Reception of Darwinism in Uruguay," in *The Reception of Darwinism in the Iberian World: Spain, Spanish America and Brazil*, ed. Thomas F. Glick, Miguel Angel Puig-Samper, and Rosaura Ruiz (Dordrecht: Kluwer Academic Publishers, 2010), 44.
25. Mariano Soler, *La masonería y el catolicismo: Estudio comparado bajo el aspecto del derecho común, las instituciones democráticas y filantrópicas, la civilización y su influencia social* (Montevideo: Andrés Ríus, 1885), 287–89, *ia802305.us.archive.org/27/items/catolicosymasoneoosole/catolicosymasoneoosole.pdf.*
26. Soler, *La masonería y el catolicismo*, 152.
27. The speech was later published as a book entitled *Carlos Roberto Darwin*. Eduardo Ladislao Holmberg, *Carlos Roberto Darwin* (Buenos Aires: El Nacional, 1882), 17, *books.google.com/books/about/Carlos_Roberto_Darwin.html?id=*UJ0aAAAAYAAJ.

28. Domingo F. Sarmiento repeatedly nationalized Darwin, stressing the role the Argentine turf played in the development of his ideas. Throughout his work, Holmberg continually and intentionally illustrated Darwin's arguments with examples from the River Plate. Holmberg is thus not merely applying Darwin to the River Plate context, but rather claiming that Darwinian-style evolution as a theory emerged from River Plate lands and histories. That Darwin wrote and attached his name to the theory is only a matter of chance, for the universality of the laws meant they could have been "discovered" anywhere.
29. Glick, "The Reception of Darwinism in Uruguay," 45.
30. Juan Zorrilla de San Martín, "Darwin," in *Juan Zorrilla de San Martín en la prensa: Escritos y discursos*, ed. Antonio Seluja Cecín (Montevideo: Comisión Nacional de Homenaje del Sesquicentenario de los Hechos Históricos de 1825, 1975), 1:47.
31. Zorrilla de San Martín, "Darwin," 47.
32. Unamuno to Zorrilla de San Martín, 13 April 1909, in *Correspondencia de Zorrilla de San Martín y Unamuno*, ed. Arturo Sergio Visca (Montevideo: Instituto Nacional de Investigaciones y Archivos Literarios, 1955), 45.
33. Juan Zorrilla de San Martín, *Tabaré*, ed. Antonio Seluja Cecín (Montevideo: Universidad de la República, 1984). All *Tabaré* citations refer to line numbers.
34. Indigenous peoples' paganism was used to justify the Catholic Kings' sovereign power over the Americas, while a lack of rationality permitted their representation as natural slaves. Anthony Pagden, *The Fall of Natural Man: The American Indian and the Origins of Comparative Ethnology* (Cambridge: Cambridge University Press, 1982), 29, 43.
35. Robert Knox, *The Races of Men: A Philosophical Enquiry into the Influence of Race over the Destinies of Nations*, 2nd ed. (London: Henry Renshaw, 1862), 66–67.
36. de Gobineau, *The Inequality of Human Races*, 64.
37. Antonio Seluja Cecín and Alberto Paganini, *"Tabaré": Proceso de creación* (Montevideo: Biblioteca Nacional, 1979), 94.
38. In many ways, *Tabaré* is a prime example of what Werner Sollors has called the "warring blood melodrama." Common in US abolitionist literature, these texts attributed a character's psychology and actions to their racial traits. Werner Sollors, *Neither Black nor White yet Both: Thematic Explorations of Interracial Literature* (Oxford: Oxford University Press, 1997), 243. A common trope was the *tragic mulatto* for whom being of mixed race caused physical problems and severe psychological torment leading to a tragic ending. Sterling Brown, "Negro Character as Seen by White Authors," *Journal of Negro Education* 2, no. 2 (April 1933): 193–95. Tabaré fits this mold, while the emphasis on physicality and emotion instead of reason suggests that miscegenation is yet another mechanism of degeneration, reducing the mestizo from fully human to beast-like savage.
39. Broca, *On the Phenomenon of Hybridity*.
40. Young, *Colonial Desire*, 16. Nancy Stepan argues that the belief that "improper racial mixing resulted in degeneracy" was extremely persistent, surviving for many years despite other changes in scientific knowledge. Nancy Stepan, "Biological Degenera-

tion: Races and Proper Places," in *Degeneration: The Dark Side of Progress*, ed. J. Edward Chamberlin and Sander L. Gilman (New York: Columba University Press, 1985), 105.

41. Rubén Dellarciprete, "Permanencia y superación del '80 en dos escritores de 'entre siglos': Enrique Loncán y Eduardo L. Holmberg" (PhD diss, Universidad Nacional de La Plata, 2011), 24, *sedici.unlp.edu.ar/handle/10915/30897*.

42. de las Casas, *Brevísima relación*, 81, 162.

43. Operé, *Indian Captivity in Spanish America*, 220.

44. Holmberg, *Carlos Roberto Darwin*, 65-66.

45. Holmberg, *Carlos Roberto Darwin*, 67. Throughout the speech, Holmberg's understanding of Darwinism draws heavily from Ernst Haeckel's 1868 *Naturliche Schopfungsgeschicte* (translated to English in 1876, and to Spanish in 1879 by Claudio Curveir González), as evidenced by these passages.

46. Novoa and Levine, *From Man to Ape*, 122.

47. Juan Zorrilla de San Martín. "A mi América española," in *Huerto Cerrado* (Montevideo: Imprenta Nacional Colorada, 1930), 319.

48. Daniel Garrison Brinton, *Races and Peoples: Lectures on the Science of Ethnography* (New York: N. D. C. Hodges, 1890), 287.

49. Juan Zorrilla de San Martín, "Descubrimiento y conquista del Río de la Plata," in *Juan Zorrilla de San Martín: Conferencias y discursos*, ed. Benjamín Fernández y Medina (Montevideo: Imprenta Nacional Colorada, 1930), 1:59.

50. James Cowles Prichard, *The Natural History of Man: Comprising Inquiries into the Modifying Influence of Physical and Moral Agencies on the Different Tribes of the Human Family* (London: H. Bailliere, 1843), 427.

51. Novoa and Levine, *From Man to Ape*, 62.

52. Holmberg's original Spanish also speaks of the *germen* (translated here as *seed*), another word which entered the vocabulary in the nineteenth century. According to the NTLLE, *germen* first appeared in a dictionary in 1803. *Gestación* appeared for the first time in 1852. Real Academia Española, *Nuevo tesoro lexicográfico de la lengua española*, s.v. "germen," s.v. "gestación," *ntlle.rae.es/ntlle/SrvltGUIMenuNtlle?cmd=Lema&sec=1.0.0.0.0.*, accessed May 10, 2018.

53. Novoa and Levine, *From Man to Ape*, 76.

54. Novoa, "The Rise and Fall," 184.

55. Young, *Colonial Desire*, 9.

56. M. C. Bobillo et al., "Amerindian Mitochondrial DNA Haplogroups Predominate in the Population of Argentina: Towards a First Nationwide Forensic Mitochondrial DNA Sequence Database," *International Journal of Legal Medicine* 123, no. 4 (July 2010): 263–68, *www.ncbi.nlm.nih.gov/pubmed/19680675*.

57. Mónica Sans, "'Raza,' adscripción étnica y genética en Uruguay," *Runa* 30, no. 2 (2009): 167.

58. "Lin-Calél. Del poema así titulado, próximo a aparecer," *Caras y Caretas* 14, no. 676 (16 Sept. 1911).

59. Dellarciprete, "Permanencia y superación del '80," 27.

CHAPTER 5

1. Eduarda Mansilla de García, *Lucía Miranda* (1860), ed. María Rosa Lojo (Madrid: Iberoamericana, 2007); Eduarda Mansilla de García, *Recuerdos de viaje*, ed. J. P. Spicer-Escalante (Buenos Aires: Stockcero, 2006).
2. Florence Dixie, *Across Patagonia* (London: Richard Bentley and Son, 1880).
3. Florence Dixie, *The Two Castaways: Adventures in Patagonia* (London: John F. Shaw, 1890); Florence Dixie, *Aniwee, or, The Warrior Queen: A Tale of the Araucano Indians and the Mythical Trauco People* (London: Henry and Co., 1890).
4. The first cohort of women at university finished their degrees around 1910. During this period, they were actively publishing scientific works and beginning jobs as professors, museum assistants, and researchers. The Museo de La Plata was a particularly accessible place for female scholars, and García notes women participating in classes, discussions, conference planning, and other activities. Susana V. García, "Mujeres, ciencias naturales y empleo académico en la Argentina (1900–1940)," INTER*thesis* 8, no. 2 (July–December 2011): 90–91, DOI: 10.50071807-1384.2011v8n2p83. Already in 1911, Carolina Etile Spegazzini had graduated in natural sciences and teaching and had published her first article (with a female co-author, María Luisa Cobanera) in the *Revista del Museo de La Plata.* (García, "Mujeres," 95). The VII Congreso Internacional de Americanistas in Buenos Aires in 1910 saw the formal participation of three Argentine female anthropologists. Dora Barrancos, "Itinerarios científicos femeninos a principios de siglo XX: solas, pero no resignadas," in *La ciencia en la Argentina entre siglos: Textos, contextos, instituciones*, ed. Marcelo Montserrat (Buenos Aires: Manantial, 2000), 127.
5. Natália Fontes de Oliveira, *Three Traveling Women Writers: Cross-Cultural Perspectives of Brazil, Patagonia, and the US from the Nineteenth Century* (New York: Routledge, 2017), 9.
6. García, "Mujeres," 91.
7. Ana Carolina Arias, "Las mujeres en la historia de la ciencia argentina: Una revisión crítica de la bibliografía," *Trabajos y Comunicaciones, 2da Época* 43 (2016): 8.
8. Masiello, *Between Civilization*, 19.
9. Frederick, *Wily Modesty*, 86.
10. "Homenaje a Darwin: Sarmiento y Holmberg," *El Nacional*, May 20, 1882.
11. Frederick, *Wily Modesty*, 80–95. The closest comparison to Mansilla's work would be the other woman-authored versions of the Lucía Miranda myth: Rosa Guerra's *Lucía Miranda* (1860) and Celestina Funes's *Lucía Miranda: Episodio nacional* (1883). These texts are less interesting to me in this work, for neither Guerra nor Funes had the same connections to the scientific-political elite that Mansilla did and neither of their texts depict indigenous women in any significant way.
12. Marina L. Guidotti, "Eduarda Mansilla, periodismo y exaltación del ser nacional," *Signos Universitarios* 53 (2017): 121. *Lucía Miranda* was first published as a *folletín* in *La Tribuna*, the same paper where Lucio would publish *Excursión* ten years later. It is also the same paper where José Manuel Estrada published his attack on evolutionary thought in 1862. Mansilla's own journalistic work has been carefully

compiled by Guidotti. Mansilla describes being caught reading Comte by her intellectual mentor, Jacobo Bermúdez de Castro, who exclaimed, "That rationalism is a shameful materialism!" "Filósofo y poeta (conclusión)," *El Nacional*, May 11, 1881, in Eduarda de García Mansilla, *Escritos periodísticas completos, 1860–1892: Eduarda Mansilla de García*, ed. Marina Guidotti (Buenos Aires: Corregidor, 2015), 479. The references to Darwin are found in "Siempre sobre Moda," *El Nacional*, December 20, 1880, in *Escritos periodísticas completos*, 407 and "Opinión de una dama en la cuestión religiosa: Ser o no ser," *El Nacional*, May 5, 1884, *Escritos periodísticas completos*, 645.

13. Fontes de Oliveira, *Three Traveling Women Writers*, 84.

14. Eduarda Mansilla, "Opinión de una dama," 644.

15. Eduarda Mansilla, "Opinión de una dama," 645.

16. Mónica Szurmuk, *Women in Argentina: Early Travel Narratives* (Gainesville: University Press of Florida, 2000), 69.

17. Fiona J. Mackintosh, "Travellers' Tropes: Lady Florence Dixie and the Penetration of Patagonia," in *Patagonia: Myths and Realities*, ed. Fernanda Peñaloza, Jason Wilson, and Claudio Canaparo (Oxford: Peter Lang, 2011), 76.

18. Jawad Syed and Faiza Ali, "The White Woman's Burden: From Colonial *Civilisation* to Third World *Development*," *Third World Quarterly* 32, no. 2 (March 2011): 355–56, DOI: 10.1080/01436597.2011.560473.

19. Mabel Moraña, Enrique Dussel, and Carlos A. Jáuregui, "Colonialism and Its Replicants," in *Coloniality at Large: Latin America and the Postcolonial Debate*, ed. Mabel Moraña, Enrique Dussel, and Carlos A. Jáuregui (Durham: Duke University Press, 2008), 5.

20. Fontes de Oliveira, *Three Traveling Women Writers*, 13.

21. For more on Mansilla's life, see María Rosa Lojo, introduction to *Lucía Miranda*, by Eduarda Mansilla, ed. María Rosa Lojo (Madrid: Iberoamericana, 2007), 12–19.

22. Tracing the Lucía Miranda legend's many versions is a popular critical pastime. See, for example, Nancy Hanway, "Valuable White Property: Lucía Miranda and National Space," *Chasqui* 30, no. 1 (May 2001): 115–30; Kathryn Lehman, "Naturaleza y cuerpo femenino en dos narrativas argentinas de origen nacional," *Revista Iberoamericana* 70, no. 206 (March 2004): 117–24; María Rosa Lojo, introduction to *Lucía Miranda*, 25–54; Francine Masiello, *Between Civilization and Barbarism*, 37; and Susana Rotker, "*Lucía Miranda*: Negación y violencia del origen," *Revista Iberoamericana* 58, no. 178–179 (June 1997): 115–27.

23. In Argentina, the settlement is usually referred to as Sancti Spiritu and the river as Carcarañá; I use Mansilla's spelling variants.

24. Frederick, *Wily Modesty*, 88; Masiello, *Between Civilization and Barbarism*, 43; Rotker, "Lucía Miranda," 124–25; Vania Barraza Toledo, "*Lucía Miranda*, de Eduarda Mansilla: La española (que) cautiva en América," in *El cautiverio en la literatura del Nuevo Mundo*, ed. Miguel Donoso Pareja, Mariela Insúa, and Carlos Mata Induráin (Madrid: Iberoamericana, 2011), 34.

25. Hanway, *Embodying Argentina*, 121.

26. María Laura Pérez Gras, "Ojos visionarios y voces transgresoras: La cuestión

del Otro en los relatos de viajes de los hermanos Mansilla," *Anales de la literatura hispanoamericana* 39 (2010): 300.

27. Pérez Gras, "Ojos visionarios," 300.

28. Novoa, "Unclaimed Fright," 551.

29. Hanway, "Valuable White Property," 127.

30. Remedios Mataix, "Romanticismo, feminidad e imaginarios nacionales: Las *Lucía Miranda* de Rosa Guerra y Eduarda Mansilla," *Río de La Plata* 29–30 (2006): 220–21.

31. Masiello, *Between Civilization*, 77.

32. One possible intertext of this idea is Louis-Aimé Martin's *De l'éducation des mères de famille, ou, De la civilisation du genre humain par les femmes* (1834). In an 1883 article in *La Nación*, Mansilla cites the Spanish translation of Martin's work, which was first published in Barcelona in 1849. It is thus possible that Mansilla may have read it prior to writing *Lucía Miranda*. Eduarda Mansilla, "Educación de la mujer," *La Nación*, July 28, 1883, in *Escritos periódicos completos*, 621.

33. Masiello, *Between Civilization and Barbarism*, 43.

34. Lojo, introduction to *Lucía Míranda*, 20–21.

35. Mataix, "Romanticismo," 223.

36. Novoa, "Unclaimed Fright," 553.

37. Olena Shkatulo, "Nation, Gender, and Beyond in Mansilla's *Recuerdos de viaje*," *Journal of Iberian and Latin American Research* 23, no. 1 (2017): 67, DOI: 10.1080/13260219.2017.1308000.

38. Miguel Malarín to Julio A. Roca, 31 May 1879, sala VII legajo 7, Julio A. Roca archive, Archivo General de la Nación, Buenos Aires, Argentina.

39. Miguel Malarín to Julio A. Roca, 4 February 1879, sala VII legajo 7, Julio A. Roca archive, Archivo General de la Nación, Buenos Aires, Argentina.

40. James F. Lee, "The Lady Footballers and the British Press, 1895," *Critical Survey* 24, no. 1 (2012): 88.

41. Florence Dixie, *Gloriana, or, The Revolution of 1900* (London: Henry and Co., 1890), 334–36, *hdl.handle.net/2027/dul1.ark:/13960/t7br9fz74*.

42. The choice of *Inacayal* for the villain's name raises the question of if Dixie had heard of the real cacique Inacayal—later resident of Moreno's museum—while traveling in Patagonia. Given the areas she visited and the Tehuelches she met, it is entirely possible.

43. Joseph Bristow, *Empire Boys: Adventures in a Man's World* (London: HarperCollins, 1991), 15.

44. Bristow, *Empire Boys*, 20.

45. Jeffrey Richards, introduction to *Imperialism and Juvenile Literature*, ed. Jeffrey Richards (Manchester: Manchester University Press, 1989), 2. As Sara Mills argues, writing about imperialism was "as much about constructing *masculine* British identity as constructing a national identity per se." Sara Mills, *Discourses of Difference: An Analysis of Women's Travel Writing and Colonialism* (London: Routledge, 1993), 3.

46. Michelle Smith, *Empire in British Girls' Literature and Culture: Imperial Girls, 1880–1915* (New York: Palgrave/Macmillan, 2011), 15.

47. Musters, *At Home with the Patagonians*, 126; Dixie, *Aniwee*, 64.

48. Lila Marz Harper, *Solitary Travelers: Nineteenth-Century Women's Travel Narratives and the Scientific Vocation* (Madison: Fairleigh Dickinson University Press, 2001), 136.

49. Of course, the marginalization of indigenous peoples as "no one" has a long history in Argentine and international writing about Patagonia.

50. Simon Gikandi, *Maps of Englishness: Writing Identity in the Culture of Colonialism* (New York: Columbia University Press, 1996), 122.

51. Dixie, *Across Patagonia*, 68.

52. Lista, *Viaje al país de los tehuelches*, 90.

53. Miguel Malarín to Julio A. Roca, 31 May 1879, sala VII legajo 7, Julio A. Roca archive, Archivo General de la Nación, Buenos Aires, Argentina.

54. Mackintosh, "Travellers' Tropes," 84. Mónica Szurmuk, in contrast, sees positive representations of indigenous society even in Dixie's travelogue. Szurmuk claims that in *Across Patagonia*, Dixie used Tehuelche life to comment on British gender structures, arguing that the Tehuelches' stable marriages and affectionate approach to child rearing were preferable to British infidelity and parental distance (Szurmuk, *Women in Argentina*, 75). Given its basis on one short paragraph, I find this conclusion weak and am more inclined to agree with Mackintosh's assessment.

55. Monica Anderson, *Women and the Politics of Travel, 1870–1914* (Madison: Fairleigh Dickinson University Press, 2006), 139.

56. H. S. Ferns, *Britain and Argentina in the Nineteenth Century* (New York: Arno Press, 1977), 112–13.

57. Ferns, *Britain and Argentina*, 397.

58. Daniel Schávelzon, "Argentina and Great Britain: Studying an Asymmetrical Relationship through Domestic Material Culture," *Historical Archaeology* 47, no. 1 (2013): 10, *www.jstor.org/stable/43491296*.

59. Robert D. Aguirre, *Informal Empire: Mexico and Central America in Victorian Culture* (Minneapolis: University of Minnesota Press, 2005), xv.

60. Aguirre, *Informal Empire*, xvi.

61. Anderson, *Women*, 29.

62. Antoinette Burton, "The White Woman's Burden: British Feminists and 'The Indian Woman,' 1865–1915," in *Western Woman and Imperialism: Complicity and Resistance*, ed. Nupur Chaudhuri and Margaret Strobel (Bloomington: Indiana University Press, 1992), 137.

63. Musters, *At Home with the Patagonians*, 96.

64. Musters, *At Home with the Patagonians*, 30.

65. Moreno, *Viaje a la Patagonia austral*, 455.

66. Lista, *Viaje al país de los tehuelches*, 65.

67. J. S. Bratton, "British Imperialism and the Reproduction of Femininity in Girls' Fiction, 1900–1930," in *Imperialism and Juvenile Literature*, ed. Jeffrey Richards (Manchester: Manchester University Press, 1989), 196.

68. Zeballos, *Descripción amena*, 386.

69. Zeballos, *Descripción amena*, 390.

CONCLUSION

1. Qtd. in Mases, *Estado y cuestión indígena*, 59.
2. Alfredo Ebelot, *La pampa: Costumbres argentinas*, ed. José Roberto del Río (Buenos Aires: Ciordia & Rodríguez, 1952), 16.
3. Cesare Lombroso, "Señor Profesor Clemente Onelli," *Revista del Jardín Zoológico de Buenos Aires* Epóca II, 4 (1908): 203.
4. Onelli wrote about indigenous alcoholism and their general decline in *Trepando los Andes: Un naturalista en la Patagonia argentina*, ed. Carlos Fernández Balboa (Buenos Aires: Continente, 2007), 69–78. Lucio Mansilla and Lista also highlighted the negative effects of alcohol in their works.
5. Stein, *Measuring Manhood*, 20. Italics in original.
6. Rodríguez, *Civilizing Argentina*, 33-34.
7. Lucas Ayarragary, "La constitución étnica argentina y sus problemas," *Archivos de psiquiatría y criminología aplicadas a las ciencias afines* 11 (1912): 23. *babel.hathitrust.org/cgi/pt?id=mdp.35112103098788*.
8. Stepan,"*The Hour of Eugenics*," 103.
9. Rodríguez, *Civilizing Argentina*, 7.
10. Nari, *Políticas de maternidad*, 46.
11. Endere, "Cacique Inakayal," 2.
12. María Luz Endere, "The Reburial Issue in Argentina: A Growing Conflict," in *The Dead and Their Possessions: Repatriation in Principle, Policy, and Practice*, ed. Cressida Fforde, Jane Hubert, and Paul Turnball (London: Routledge, 2002), 268.
13. Florencia Bonelli, *Indias blancas* (Buenos Aires: Punto de lectura, 2012); Florencia Bonelli, *Indias blancas: La vuelta del ranquel*; and Gloria Casañas, *La maestra de la laguna: Un amor entre Boston y las pampas* (Buenos Aires: Plaza Janes, Sudamericana, 2011).
14. Ashley Elizabeth Kerr, "Indigenous Lovers and Villainous Scientists: Rewriting Nineteenth-Century Ideas of Race in Argentine Romance Novels," *Chasqui* 48, no. 1 (May 2019): 293–310.
15. Myriam Angueira and Guillermo Glass, *Inacayal: La negación de nuestra identidad,* (INCAA/FronteraCine, 2012), DVD.
16. Lojo writes about her experiences in María Rosa Lojo and Carlos Mayol Laferrere, *Tras las huellas de Mansilla: Contexto histórico y aportes críticos a* Una excursión a los indios ranqueles (Córdoba: La Copista, 2012).
17. María Rosa Lojo, *La pasión de los nómades* (Buenos Aires: Debolsillo, 2008).
18. María Rosa Lojo, *Una mujer de fin de siglo* (Buenos Aires: Stockcero, 2007); María Rosa Lojo, *Historias ocultas en la Recoleta* (Buenos Aires: Suma de Letras, 2002).
19. Novels about the Fuegians include Sylvia Iparraguirre, *La tierra del fuego* (Buenos Aires, Alfaguara, 1998) and Eduardo Belgrano Rawson, *Fuegía* (Buenos Aires: Sudamericana, 2002). Present-day texts about Onelli include Leopoldo Brizuela, *La locura de Onelli* (Buenos Aires: Bajo la luna, 2012) and Alberto Perrone, *La jirafa de Clemente Onelli* (Buenos Aires: Sudamericana, 2012).
20. Silvia Ametrano, "20 años después: restitución complementaria del Cacique Inacayal y familiares," *Museo* 27 (2014-2015): 10, *sedici.unlp.edu.ar/bitstream/handle/10915/52084/Documento_completo.pdf?sequence=1*.

BIBLIOGRAPHY

Aguirre, Robert D. *Informal Empire: Mexico and Central America in Victorian Culture.* Minneapolis: University of Minnesota Press, 2005.

Alberdi, Juan Batista. *Bases y puntos de partida para la organización política de la República Argentina.* Buenos Aires: La Cultura Argentina, 1928. *archive.org/download/basesypuntosdepa00albe/basesypuntosdepa00albe.pdf.*

Alonso, Carlos. "Oedipus in the Pampas: Lucio Mansilla's *Una Excursión a Los Indios Ranqueles.*" *Revista de estudios hispánicos* 24, no. 2 (May 1990): 39–59.

Alvarado Pérez, Margarita. "Pose y montaje en la fotografía *mapuche*: Retrato fotográfico, representación e identidad." In *Mapuche fotografías siglos XIX y XX: Construcción y montaje de un imaginario*, edited by Margarita Alvarado Pérez, Pedro Mege, and Christian Báez, 13–27. Santiago: Pehuén, 2001.

Alvarado Pérez, Margarita, Pedro Mege, and Christian Báez, eds. *Mapuche fotografías siglos XIX y XX: Construcción y montaje de un imaginario.* Santiago: Pehuén, 2001.

Ameghino, Florentino. *Obras completas y correspondencia científica de Florentino Ameghino.* Vol. 20, *Correspondencia Científica*, edited by Alfredo J. Torcelli. La Plata: Taller de Impresiones Oficiales, 1935.

Ametrano, Silvia. "20 años después: restitución complementaria del Cacique Inacayal y familiares." *Museo* 27 (2014-2015): 5-10, *sedici.unlp.edu.ar/bitstream/handle/10915/52084/Documento_completo.pdf?sequence=1.*

______. "Los procesos de restitución en el Museo de La Plata." *Revista Argentina de Antropología Biológica* 17, no. 2 (Dec. 2015): n.p.

Andermann, Jens. *Mapas de poder: Una arqueología literaria del espacio argentino.* Rosario: Beatriz Viterbo, 2000.

______. *The Optic of the State: Visuality and Power in Argentina and Brazil.* Pittsburgh: University of Pittsburgh Press, 2007.

Anderson, Mónica. *Women and the Politics of Travel, 1870–1914.* Madison: Fairleigh Dickinson University Press, 2006.

Andrews, George Reid. *The Afro-Argentines of Buenos Aires, 1800–1900.* Madison: University of Wisconsin Press, 1980.

Angueira, Myriam, and Guillermo Glass. *Inacayal: La negación de nuestra identidad.* INCAA/FronteraCine, 2012. DVD.

Appelbaum, Nancy P., Anne S. Macpherson, and Karin Alejandra Rosemblatt. Introduction to *Race and Nation in Modern Latin America*, edited by Nancy P. Appelbaum, Anne S. Macpherson, and Karin Alejandra Rosemblatt, 1–31. Chapel Hill: University of North Carolina Press, 2003.

Arcos, Santiago. "Cuestión de indios." In *Cuestión de indios*, edited by Juan Bautista Sasiaiñ, 7–30. Buenos Aires: Policía Federal Argentina, 1979.

______. *La Plata: Etude historiqué.* Paris: Michel Levy Freres, 1865.

Argentine Republic. "Constitución de la Confederación Argentina (May 1, 1853)." WIPO Lex No. AR147, *www.wipo.int/wipolex/en/text.jsp?file_id=361080.*

Argentina, Cámara de Diputados. *Diario de sesiones de la Cámara de Diputados: Año 1885.* 2 vols. Buenos Aires: Moreno y Nuñez, 1886.

______. *Diario de sesiones de la Cámara de Diputados: Año 1888*, vol. 1. Buenos Aires: Sudamericana, 1889.

Argeri, María E. *De guerreros a delincuentes: La desarticulación de las jefaturas indígenas y el poder judicial. Norpatagonia, 1880–1930.* Madrid: Consejo Superior de Investigaciones Científicas, 2005.

Arias, Ana Carolina. "Las mujeres en la historia de la ciencia argentina: Una revisión crítica de la bibliografía." *Trabajos y Comunicaciones, 2da Época* 43 (2016): 1–16.

Augustine-Adams, Kif. "'She Consents Implicitly': Women's Citizenship, Marriage, and Liberal Political Theory in Late-Nineteenth- and Early-Twentieth-Century Argentina." *Journal of Women's History* 13, no. 4 (2002): 8–30.

Avedaño, Santiago. *Memorias del ex-cautivo Santiago Avedaño.* Edited by P. Meinrado Hux. Buenos Aires: Elefante Blanco, 1999.

______. *Usos y costumbres de los indios de la Pampa.* Edited by P. Meinrado Hux. Buenos Aires: Elefante Blanco, 2000.

Ayarragary, Lucas. "La constitución étnica argentina y sus problemas." *Archivos de psiquiatría y criminología aplicadas a las ciencias afines* 11 (1912): 22–42. *babel.hathitrust.org/cgi/pt?id=mdp.35112103098788.*

Azar, Pablo, Gabriela Nacach, and Pedro Navarro Floria. "Antropología, genocidio y olvido en la representación del otro étnico a partir de la Conquista." In *Paisajes del progreso: La resignificación de la Patagonia Norte, 1880–1916*, edited by Pedro Navarro Floria, 79–106. Neuquen: Educo, 2007.

Babini, José. *La ciencia en la Argentina.* Buenos Aires: Eudeba, 1963.

Bacigalupo, Ana Mariella. *Shamans of the Foye Tree: Gender, Power, and Healing among Chilean Mapuche.* Austin: University of Texas Press, 2007.

Baigorria, Manuel. *Memorias.* Edited by P. Meinrado Hux. Buenos Aires: Elefante Blanco, 2006.

Balderston, Daniel. "The Indianist Novels of Estanislao S. Zeballos." *Revista Canadiense de Estudios Hispánicos* 15, no. 2 (Winter 1991): 323–27. *www.jstor.org/stable/27762828.*

Barrancos, Dora. "Itinerarios científicos femeninos a principios de siglo XX: Solas, pero no resignadas." In *La ciencia en la Argentina entre siglos: Textos, contextos, instituciones*, edited by Marcelo Montserrat, 127–44. Buenos Aires: Manantial, 2000.

Barraza Toledo, Vania. "*Lucía Miranda*, de Eduarda Mansilla: La española (que) cautiva en América." In *El cautiverio en la literatura del Nuevo Mundo*, edited by Miguel Donoso Pareja, Mariela Insúa, and Carlos Mata Induráin, 27–40. Madrid: Iberoamericana, 2011.

Bechis, Martha. "Rostros aborígenes de las pampas argentinas, siglos XVIII–XIX." *Boletín* TEFROS 2, no. 2 (Winter 2004): n.p. *www.hum.unrc.edu.ar/ojs/index.php/tefros/article/view/78/61*.

Beer, Gillian. *Open Fields: Science in Cultural Encounter*. Oxford: Oxford University Press, 1996.

Belgrano Rawson, Eduardo. *Fuegía*. Buenos Aires: Sudamericana, 2002.

Ben, Pablo. "Male Same-Sex Sexuality and the Argentine State, 1880–1930." In *The Politics of Sexuality in Latin America*, edited by Javier Corrales and Mario Pecheny, 33–43. Pittsburgh: University of Pittsburgh Press, 2010.

Berco, Cristian. "Silencing the Unmentionable: Non-Reproductive Sex and the Creation of a Civilized Argentina, 1860–1900." *The Americas* 58, no. 3 (2002): 419–41.

Biagini, Hugo. *Cómo fue la generación del 80*. Buenos Aires: Plus Ultra, 1980.

Bieder, Robert. *Science Encounters the Indian, 1820–1880: The Early Years of American Ethnology*. Norman: University of Oklahoma Press, 1986.

Bjerg, María. "Vínculos mestizos: Historias de amor y parentesco en la campaña de Buenos Aires en el siglo XIX." *Boletín del Instituto de Historia Argentina y Americana "Dr. Emilio Ravignani,"* 3rd ser., 30 (2007): 73–99.

Bloom, John. *To Show What an Indian Can Do: Sports at Native American Boarding Schools*. Minneapolis: University of Minnesota Press, 2000.

Bobillo, M. C., B. Zimmermann, A. Sala, G. Huber, H. J. Bandelt, D. Corach, and W. Parson. "Amerindian Mitochondrial DNA Haplogroups Predominate in the Population of Argentina: Towards a First Nationwide Forensic Mitochondrial DNA Sequence Database." *International Journal of Legal Medicine* 123, no. 4 (July 2010): 263–68. *www.ncbi.nlm.nih.gov/pubmed/19680675*.

Bonelli, Florencia. *Indias blancas*. Buenos Aires: Punto de lectura, 2012.

———. *Indias blancas: La vuelta del ranquel*. Buenos Aires: Suma de letras, 2012.

Bratton, J. S. "British Imperialism and the Reproduction of Femininity in Girls' Fiction, 1900–1930." In *Imperialism and Juvenile Literature*, edited by Jeffrey Richards, 195–215. Manchester: Manchester University Press, 1989.

Brickhouse, Anna. *The Unsettlement of America: Translation, Interpretation, and the Story of Don Luis de Velasco, 1560–1945*. Oxford: Oxford University Press, 2015.

Brinton, Daniel Garrison. *Races and Peoples: Lectures on the Science of Ethnography*. New York: N. D. C. Hodges, 1890.

Briones, Claudia, and José Luis Lanata. "Living on the Edge." In *Archaeological and Anthropological Perspectives on the Native Peoples of Pampa, Patagonia and Tierra del Fuego to the Nineteenth Century*, edited by Claudia Briones and José Luis Lanata, 1–12. Westport: Bergin & Garvey, 2002.

Bristow, Joseph. *Empire Boys: Adventures in a Man's World*. London: HarperCollins, 1991.

Brizuela, Leopoldo. *La locura de Onelli*. Buenos Aires: Bajo la luna, 2012.

Broca, Paul. *On the Phenomena of Hybridity in the Genus Homo*. Edited and translated by C. Carter Blake. London: Longman, Green, Longman, and Roberts, 1864.

Brown, J. Andrew. *Test Tube Envy: Science and Power in Argentine Narrative*. Lewisburg: Bucknell University Press, 2005.

Brown, Sterling. "Negro Character as Seen by White Authors." *Journal of Negro Education* 2, no. 2 (April 1933): 179–203.

John Browning, "Cornelius de Pauw and Exiled Jesuits: The Development of Nationalism in Spanish America," *Eighteenth-Century Studies* 11, no. 3 (1978): 289–307, *www.jstor.org/stable/2738194*.

Burton, Antoinette. "The White Woman's Burden: British Feminists and 'The Indian Woman,' 1865–1915." In *Western Woman and Imperialism: Complicity and Resistance*, edited by Nupur Chaudhuri and Margaret Strobel, 137–57. Bloomington: Indiana University Press, 1992.

Bustos, Jorge, and Leonardo Dam. "El Registro de Vecindad del Partido de Patagones (1887) y los niños indígenas como botín de guerra." *Corpus: Archivo virtual de la alteridad americana* 2, no. 1 (2012): 1–4. *ppct.caicyt.gov.ar/index.php/corpus/article/view/1413*.

Butto, Ana. "Artefactos autóctonos y foráneos en las fotografías de indígenas y criollos militares durante la 'Conquista Del Desierto' (Norpatagonia, Siglo XIX)." In *Tendencias teórico-metodológicas y casos de estudio en la arqueología de la Patagonia*, edited by Atilio Zangrando, Ramiro Barberena, Adolfo Gil, Gustavo Neme, Miguel Giardina, Leandro Luna, Clara Otaola, Salvador Paulides, Laura Salgán, and Angela Tívoli, 53–62. San Rafael: Museo de Historia Natural de San Rafael, 2013.

______. "Con el foco en el otro: Las representaciones visuales acerca del indio y el territorio en los expedicionarios de la Conquista Del Desierto en las campañas de 1879 y 1883." In *Entre pasados y presentes III: Estudios contemporáneos en ciencias antropológicas*, edited by Nora Kuperszmit, Teresa Lagos Mármol, Leonardo Mucciolo, and Marian Sacchi, 105–21. Buenos Aires: Mnemosyne, 2012.

______. "Visualización de los roles de género en las fotografías etnográficas de mapuches y tehuelches (siglox XIX-XX)." *RiHumSo* 7, no. 13 (November 2018): 21–46.

Butto, Ana R., and Dánae Fiore. "Violencia fotografiada y fotografías violentas: Acciones agresivas y coercitivas en las fotografías etnográficas de pueblos originarios fueguino y patagónicos." *Nuevo Mundo Mundos Nuevos* (2014): 1–24.

Caggiano, Sergio. "La visión de la 'raza': Apuntes para un estudio de la fotografía de tipos raciales en Argentina." *Revista del Museo de Antropología* 6 (2013): 107–18.

Campanella, Hebe Noemí. *La generación del 80: Su influencia en la vida cultural argentina*. Buenos Aires: Tekné, 1983.

Carman, Jorge. Introduction to *Obras*, vol. 1, by Ramón Lista, 13–20. Edited by Jorge Carman. Buenos Aires: Confluencia, 1998.

Casamiquela, Rodolfo. "Indígenas patagónicos en el museo: Primera parte." *Revista museo* II, no. 12 (1998): 69–75.

Casañas, Gloria. *La maestra de la laguna: Un amor entre Boston y las pampas*. Buenos Aires: Plaza Janes, Sudamericana, 2011.

Castillo Bernal, María Florencia, and Liliana E. María Videla. "Estudio comparativo de tres jefaturas femeninas en Patagonia." In *Actas del VI Congreso de Historia Social y Política de La Patagonia Argentino-Chilena*, 15–18. Trevelín, Argentina: Secretaría de Cultura de la Provincia del Chubut, 2009.

Castro-Klarén, Sara, and John Charles Chasteen, eds. *Beyond Imagined Communities: Reading and Writing the Nation in Nineteenth-Century Latin America*. Washington, DC: Woodrow Wilson Center Press, 2003.

Centeno, Pablo Andrés Cuadra, and María Laura Mazzoni. "La invasión inglesa y la participación popular en la reconquista y defensa de Buenos Aires 1806–1807." *Anuario del Instituto de Historia Argentina*, no. 11 (2011): 41–70.

Chasteen, John Charles. Introduction to *Beyond Imagined Communities: Reading and Writing the Nation in Nineteenth-Century Latin America*, ix–xxv. Edited by Sara Castro-Klarén and John Charles Chasteen. Washington, DC: Woodrow Wilson Center Press, 2003.

Cobb, Matthew. "Heredity before Genetics: A History." *Nature Reviews: Genetics*, 7 (December 2006): 953–58.

Cohan Scherer, Joanna. "The Public Faces of Sarah Winnemucca." *Cultural Anthropology* 3, no. 2 (May 1988): 178–204.

Daireaux, Emile. *Vida y costumbres en el Plata: Tomo primero, La sociedad argentina.* Buenos Aires: Félix Lajouane, 1888. *archive.org/details/vidaycostumbresoodairgoog.*

Darwin, Charles. *The Voyage of the Beagle.* Auckland: Floating Press, 2008.

de Asúa, Miguel. *Una gloria silenciosa: Dos siglos de ciencia en la Argentina.* Buenos Aires: Libros del Zorzal, 2010.

de Gobineau, Arthur. *The Inequality of Human Races.* Translated by Adrian Collins. London: William Heinemann, 1915. *babel.hathitrust.org/cgi/pt?id=msu.31293103445569.*

de las Casas, Bartolomé. *Brevísima relación de la destruición de las Indias*, edited by André Saint-Lu. Madrid: Cátedra, 1989.

Dellarciprete, Rubén. "Permanencia y superación del '80 en dos escritores de 'entre siglos': Enrique Loncán y Eduardo L. Holmberg." PhD diss., Universidad Nacional de La Plata, 2011. *sedici.unlp.edu.ar/handle/10915/30897.*

Deloria, Philip J. *Playing Indian.* New Haven: Yale University Press, 1998.

Delrio, Walter, Diana Lenton, Marcelo Mustante, and Mariano Nagy. "Discussing Indigenous Genocide in Argentina: Past, Present, and Consequences of Argentinean State Policies toward Native Peoples." *Genocide Studies and Prevention: An International Journal* 5, no. 2 (2010): 138–59.

Díez, Beatriz. "Las *Memorias* de Santiago Avedaño y la trilogía de E. Zeballos." In *V Congreso Internacional de Letras*, edited by Américo Cristófalo, 1040–45. Buenos Aires: Facultad de Filosofía y Letras de la Universidad de Buenos Aires, 2012. *2012.cil.filo.uba.ar/sites/2012.cil.filo.uba.ar/files/0134%20DIEZ,%20BEATRIZ.pdf.*

Dixie, Florence. *Across Patagonia.* London: Richard Bentley and Son, 1880.

———. *Aniwee, or, The Warrior Queen: A Tale of the Araucano Indians and the Mythical Trauco People.* London: Henry and Co., 1890.

———. *Gloriana, or, The Revolution of 1900.* London: Henry and Co., 1890. *hdl.handle.net/2027/dul1.ark:/13960/t7br9fz74.*

———. *The Two Castaways: Adventures in Patagonia.* London: John F. Shaw, 1890.

Earle, Rebecca. *The Return of the Native: Indians and Myth-Making in Spanish America, 1810–1930.* Durham, NC: Duke University Press, 2007.

Ebelot, Alfredo. *La pampa: Costumbres argentinas.* Edited by José Roberto del Río. Buenos Aires: Ciordia & Rodríguez, 1952.

Edwards, Elizabeth. "The Image as Anthropological Document. Photographic 'Types': The Pursuit of Method." *Visual Anthropology* 3 (1990): 235–58.

———. "Material Beings: Objecthood and Ethnographic Photographs." *Visual Studies* 17, no. 1 (2002): 67–75.

Endere, María Luz. "Cacique Inakayal: La primera restitución de restos humanos ordenada por ley." *Corpus: Archivo virtual de la alteridad americana* 1, no. 1 (Primer semestre 2011): 1–7.

———. "The Reburial Issue in Argentina: A Growing Conflict." In *The Dead and Their Possessions: Repatriation in Principle, Policy, and Practice*, edited by Cressida Fforde, Jane Hubert, and Paul Turnball, 266–83. London: Routledge, 2002.

Espinosa, Antonio. *La Conquista del Desierto: Diario del capellán de la Expedición de 1879, Monseñor Antonio Espinosa, más tarde arzobispo de Buenos Aires*, edited by Bartolomé Galindez. Buenos Aires: Freeland, 1968.

Estrada, José Manuel. *El génesis de nuestra raza: Refutación de una lección del Dr. D. Gustavo Minelli sobre la misma materia*. Buenos Aires: La Bolsa, 1862.

Fabian, Johannes. *Time and the Other: How Anthropology Makes Its Object*. New York: Columbia University Press, 1983.

Farro, Máximo. *La formación del Museo de La Plata: Coleccionistas, comerciantes, estudiosos y naturalistas viajeros a fines del siglo XIX*. Rosario: Prohistoria, 2009.

———. "Imágenes de cráneos, retratos antropológicos y tipologías raciales: Los usos de las primeras colecciones de fotografías del Museo de La Plata a fines del siglo XIX." In *Los secretos de Barba Azul: Fantasías y realidades de los archivos del Museo de La Plata*, edited by Tatiana Kelly and Irina Podgorny, 69–103. Rosario: Prohistoria, 2012.

———. "Natural History Museums in Argentina, 1862–1906." *Museum History Journal* 9, no. 1 (January 2016): 121–34.

Feir, Donna. "The Long-Term Effects of Forcible Assimilation Policy: The Case of Indian Boarding Schools." *Canadian Journal of Economics* 49, no. 2 (2016): 433–80.

Fernández, Mabel. "La representación de las mujeres aborígenes en la iconografía patagónica." *Aljaba* 19 (2015): n.p.

Fernández, Silvia Mirta Beatriz. "Mansilla y los ranqueles. ¿Porqué Lucio V. Mansilla escribió *Una excursión a los indios ranqueles*?" In *Congreso Nacional de Historia Sobre la Conquista del Desierto*, vol. 4, 361–75. Buenos Aires: Academia Nacional de la Historia, 1981.

Ferns, H. S. *Britain and Argentina in the Nineteenth Century*. New York: Amo Press, 1977.

Fine, Gary Alan. "Ten Lies of Ethnography: Moral Dilemmas of Field Research," *Journal of Contemporary Ethnography* 22 (April 1993): 267–94.

Fiore, Dánae, and María Lydia Varela. *Memorias de papel: Una arqueología visual de las fotografías de pueblos originarios fueguinos*. Buenos Aires: Dunken, 2009.

Fontes de Oliveira, Natália. *Three Traveling Women Writers: Cross-Cultural Perspectives of Brazil, Patagonia, and the US from the Nineteenth Century*. New York: Routledge, 2017.

Frederick, Bonnie. "In Their Own Voice: The Women Writers of the Generación del 80 in Argentina." *Hispania* 74, no. 2 (May 1991): 282–89.

———. *Wily Modesty: Argentine Women Writers, 1860–1910*. Tempe: ASU Center for Latin American Studies Press, 1998.

Funes, Celestina. *Lucía Miranda: Episodio Nacional*. Rosario: Imprenta de "El Mensajero," 1883. *archive.org/details/lucamirandaepi00fn*.

García, Susana V. "Mujeres, ciencias naturales y empleo académico en la Argentina (1900–1940)." *INTERthesis* 8, no. 2 (July–December 2011): 83–103. DOI: 10.50071807-1384.2011v8n2p83.

García Ferrari, Mercedes, and Diego Galeano. "Police, Anthropometry, and Fingerprinting: Transnational History of Identification Systems from Rio de la Plata to Brazil." *História, Ciências, Saúde–Manguinhos* 23 (Dec. 2016): 1–24. *www.scielo.br/pdf/hcsm/v23s1/en_0104-5970-hcsm-23-s1-0171.pdf.*

Gasparri, Javier. "Mansilla: La promiscuidad de los cuerpos." In *II Coloquio internacional: Saberes contemporáneos desde la diversidad sexual: Teoría, crítica, praxis*, 99-112. Rosario: Universidad Nacional de Rosario-Centro de Estudios Interdisciplinarios, 2013.

Gerbi, Antonello. *The Dispute of the New World: The History of a Polemic, 1750–1900.* Translated by Jeremy Moyle. Pittsburgh: University of Pittsburgh Press, 2010.

Gikandi, Simon. *Maps of Englishness: Writing Identity in the Culture of Colonialism.* New York: Columbia University Press, 1996.

Gillies, Eva. Introduction to *A Visit to the Ranquel Indians.* Edited and translated by Eva Gilles, xix–xl. Lincoln: University of Nebraska Press, 1997.

Gilman, Sander L. "Black Bodies, White Bodies: Toward an Iconography of Female Sexuality in Late Nineteenth-Century Art, Medicine, and Literature," *Critical Inquiry* 12, no. 1 (Autumn 1985): 204–42.

———. *Difference and Pathology: Stereotypes of Sexuality, Race, and Madness.* Ithaca, NY: Cornell University Press, 1985.

Giordano, Mariana. *Indígenas en la Argentina: Fotografías 1860–1970.* Buenos Aires: El Artenauta, 2012.

Glick, Thomas F. "The Reception of Darwinism in Uruguay." In *The Reception of Darwinism in the Iberian World: Spain, Spanish America and Brazil*, edited by Thomas F. Glick, Miguel Angel Puig-Samper, and Rosaura Ruiz, 29–52. Dordrecht: Kluwer Academic Publishers, 2010.

Glick, Thomas F., Miguel Angel Puig-Samper, and Rosaura Ruiz, eds. *The Reception of Darwinism in the Iberian World: Spain, Spanish America and Brazil.* Dordrecht: Kluwer Academic Publishers, 2010.

Gómez, Leila. *La piedra del escándalo: Darwin en Argentina, 1845–1909.* Buenos Aires: Simurg, 2008.

Gonzalo Quiroga, Cristian. "Análisis del rol de las mujeres indígenas en los ámbitos de consenso, durante la segunda mitad del siglo XIX en Patagonia. Sugerencias para una nueva interpretación de caso." In *V Jornadas de historia social de la Patagonia*, compiled by Walter Delrio, Liliana Pierucci, Fabiana Ertola, Laura Méndez, Maximiliano Lezcano, Liliana Luseti, Inés Barelli, José Benclowicz, Alfredo Azcoitía, Susana Romaniuk y Viviana Fernández, 105–30. Río Negro: San Carlos de Bariloche, 2014. *biblioteca.clacso.edu.ar/Argentina/iidypca-unrn/20171115052701/pdf_108.pdf.*

Gordillo, Gastón, and Silvia Hirsch. "Indigenous Struggles and Contested Identities in Argentina: Histories of Invisibilization and Reemergence." *Journal of Latin American Anthropology* 8, no. 3 (2003): 4–30.

Grandin, Greg. "Can the Subaltern Be Seen? Photography and the Affects of Nationalism." *Hispanic American Historical Review* 84, no. 1 (2004): 83–111.

Guerra, Rosa. *Lucía Miranda.* Buenos Aires: Imprenta Americana, 1860. *www.cervantesvirtual.com/nd/ark:/59851/bmc80509.*

Guglielmini, Homero M. *Mansilla.* Buenos Aires: Ediciones Culturales Argentinas, 1961.

Guidotti, Marina L. "Eduarda Mansilla, periodismo y exaltación del ser nacional." *Signos Universitarios* 53 (2017): 119–45.

Gutmann, Matthew, ed. *Changing Men and Masculinities in Latin America.* Durham: Duke University Press, 2002.

______. Introduction to *Changing Men and Masculinities in Latin America*. Edited by Matthew Guttman, 1–26. Durham: Duke University Press, 2002.

Guy, Donna. "Lower-Class Families, Women, and the Law in Nineteenth-Century Argentina." *Journal of Family History* 10, no. 3 (1985): 318–31.

______. *Sex and Danger in Buenos Aires: Prostitution, Family, and Nation in Argentina*. Lincoln: University of Nebraska Press, 1991.

Haeckel, Ernst. *The History of Creation, or, The Development of the Earth and Its Inhabitants by the Action of Natural Causes*. Translated by E. Ray Lankester. Vol. 1 of 2. London: Henry S. King & Co., 1876.

Hanway, Nancy. *Embodying Argentina: Body, Space and Nation in 19th Century Narrative*. Jefferson: McFarland & Co., 2003.

______. "Valuable White Property: Lucía Miranda and National Space." *Chasqui* 30, no. 1 (May 2001): 115–30.

Harper, Lila Marz. *Solitary Travelers: Nineteenth-Century Women's Travel Narratives and the Scientific Vocation*. Madison: Fairleigh Dickinson University Press, 2001.

Hill, Ruth. "Ariana Crosses the Atlantic: An Archaeology of Aryanism in the Nineteenth Century River Plate." *Hispanic Issues On Line* 12 (2013): 92–110. *hdl.handle.net/11299/184420*.

______. "Entre lo transatlántico y lo hemisférico: Los proyectos raciales de Andrés Bello." *Revista Iberoamericana* 75, no. 228 (2009): 719–35.

Hirsch, Silvia. "La mujer indígena en la antropología argentina: Una breve reseña." In *Mujeres indígenas en la Argentina: Cuerpo, trabajo y poder*, edited by Silvia Hirsch, 15–25. Buenos Aires: Biblos, 2008.

Holdich, T. H. "Obituary: Dr. Francisco P. Moreno, Gold Medalist of the Society and Honorary Corresponding Member." *Geographical Journal* 55, no. 2 (February 1920): 156–57.

Holmberg, Eduardo Ladislao. *Carlos Roberto Darwin*. Buenos Aires: El Nacional, 1882. *books.google.com/books/about/Carlos_Roberto_Darwin.html?id=UJOaAAAAYAAJ*.

______. *Lin-Calél*. Buenos Aires: Talleres Gráficos de L. J. Rosso, 1910.

______. "La noche clásica de Walpurgis." *Anales de la Sociedad Científica Argentina* 22 (1886): 241–71.

Hubert, Jane, and Cressida Fforde. "Introduction: The Reburial Issue in the Twenty-First Century." In *The Dead and Their Possessions: Repatriation in Principle, Policy, and Practice*, edited by Cressida Fforde, Jane Hubert, and Paul Turnbull, 1–16. London: Routledge, 2002.

Iglesia, Cristina. "Mejor se duerme en la pampa: Deseo y naturaleza en *Una excursión a los indios ranqueles* de Lucio V. Mansilla." *Revista Iberoamericana* 63, no. 178-179 (June 1997): 185–92.

"Indigenous Peoples in Latin America." Infographic. CEPAL, September 22, 2014. *www.cepal.org/en/infografias/los-pueblos-indigenas-en-america-latina*.

Iparraguirre, Sylvia. *La tierra del fuego*. Buenos Aires: Alfaguara, 1998.

Jacobs, Margaret. "Maternal Colonialism: White Women and Indigenous Child Removal in the American West and Australia, 1880–1940." *Western Historical Quarterly* 36, no. 4 (2005): 453–76.

Jager, Rebecca K. *Malinche, Pocahontas, and Sacagawea: Indian Women as Cultural Intermediaries and National Symbols*. Norman: University of Oklahoma Press, 2015.

Kerr, Ashley Elizabeth. "Indigenous Lovers and Villainous Scientists: Rewriting Nineteenth-Century Ideas of Race in Argentine Romance Novels." *Chasqui* 48, no. 1 (May 2019): 293–310.

______. "Progress at What Price?: Defenses of Indigenous Peoples in Argentine Writing about Patagonia (1894-1904)." *Decimonónica* 16, no. 2 (Summer 2019): 17–33.

______. "'Somos una raza privilegiada': Anthropology, Race, and Nation in the Literature of the River Plate, 1870–2010." PhD diss., University of Virginia, 2013. ProQuest (3574693).

Knox, Robert. *The Races of Men: A Philosophical Enquiry into the Influence of Race over the Destinies of Nations*. 2nd ed. London: Henry Renshaw, 1862.

Kraus, Michael. "Of Photography and Men: Encounters with Historical Portrait and Type Photographs." In *Photography in Latin America: Images and Identities across Time and Space*, edited by Gisela Cánepa Koch and Ingrid Kummels, 33–63. Bielefeld: Transcript, 2016.

Kulick, Don, and Margaret Willson, eds. *Taboo: Sex, Identity, and Erotic Subjectivity in Anthropological Fieldwork*. London: Routledge, 1995.

Larrain, Nicanor. *Viajes en el "Villarino" a la costa sud de la República Argentina*. Buenos Aires: Juan A. Alsina, 1883.

Larson, Carolyne R. *Our Indigenous Ancestors: A Cultural History of Museums, Science, and Identity in Argentina, 1877–1943*. University Park: Pennsylvania State University Press, 2015.

Lavrin, Asunción. *Women, Feminism, and Social Change in Argentina, Chile, and Uruguay, 1890–1940*. Lincoln: University of Nebraska Press, 1995.

Lee, James F. "The Lady Footballers and the British Press, 1895." *Critical Survey* 24, no. 1 (2012): 88–101.

Lehman, Kathryn. "Naturaleza y cuerpo femenino en dos narrativas argentinas de origen nacional." *Revista Iberoamericana* 70, no. 206 (March 2004): 117–24.

Lehmann-Nitsche, Robert. "Etudes anthropologiques sur les indiens ona (groupe tshon) de la Terre de Feu." *Revista del Museo de La Plata* 23 (second part) (1916): 174–84.

Lenton, Diana Isabel. "Relaciones interétnicas: Derechos humanos y autocrítica en la Generación Del '80." In *La problemática indígena: Estudios antropológicos sobre pueblos indígenas de la Argentina*, edited by Juan Carlos Radovich and Alejandro O. Balazote, 27–65. Buenos Aires: Centro Editor de América Latina, 1992.

Lenzi, Juan Hilarión. "Ramón Lista, gobernador de Santa Cruz," *Argentina austral* 302 (December 1956).

Levine, Philippa. "States of Undress: Nakedness and the Colonial Imagination." *Victorian Studies* 50, no. 2 (Winter 2008): 189–219.

"Lin-Calél: Del poema así titulado, próximo a aparecer," *Caras y Caretas* 14, no. 676 (16 Sept. 1911).

Lista, Ramón. *La Gobernación de Santa Cruz*. In *Obras*, vol. 2, edited by Jorge Carman, 191–248. Buenos Aires: Editorial Confluencia, 1998.

______. *Los indios tehuelches: Una raza que desaparece*. In *Obras*, vol. 2, edited by Jorge Carman, 125–181. Buenos Aires: Editorial Confluencia, 1998.

______. *Mis exploraciones y descubrimientos en la Patagonia*. In *Obras*, vol. 1, edited by Jorge Carman, 117—242. Buenos Aires: Editorial Confluencia, 1998.

______. "El país de las Manzanas." In *Obras*, vol. 1, edited by Jorge Carman, 263–65. Buenos Aires: Editorial Confluencia, 1998.

______. *El Territorio de las Misiones*. In *Obras*, vol. 1, edited by Jorge Carman, 281–347. Buenos Aires: Editorial Confluencia, 1998.

______. *Viaje al país de los onas*. In *Obras*, vol. 2, edited by Jorge Carman, 16–118. Buenos Aires: Editorial Confluencia, 1998.

______. *Viaje al país de los tehuelches*. In *Obras*, vol. 1, edited by Jorge Carman, 43–93. Buenos Aires: Editorial Confluencia, 1998.

Liva, Yamila, and Teresa Laura Artieda. "Proyectos y prácticas sobre la educación de los indígenas: El caso de la misión franciscana de Laishí frente al juicio del Inspector José Elías Niklison (1901–1916)." *Historia de la educación* 15, no. 1 (2014): 45–68.

Lojo, María Rosa. *Historias ocultas en la Recoleta*. Buenos Aires: Suma de Letras, 2002.

______. "El indio como 'prójimo,' la mujer como el 'otro' en *Una excursión a los indios ranqueles* de Lucio V. Mansilla," *Alba de América* 26-27 (July 1996): 131–37.

______. Introduction to *Lucía Miranda*, by Eduarda Mansilla, edited by María Rosa Lojo, 11–137. Madrid: Iberoamericana, 2007.

______. *La pasión de los nómades*. Buenos Aires: Debolsillo, 2008.

______. *Una mujer de fin de siglo*. Buenos Aires: Stockcero, 2007.

Lojo, María Rosa, and Carlos Mayol Laferrere. *Tras las huellas de Mansilla: Contexto histórico y aportes críticos a* Una excursión a los indios ranqueles. Córdoba: La Copista, 2012.

Lombroso, Cesare. "Señor Profesor Clemente Onelli." *Revista del Jardín Zoológico de Buenos Aires* Epóca II 4 (1908): 203–4.

Lubbock, John. *The Origin of Civilisation and the Primitive Condition of Man: Mental and Social Condition of Savages*. New York: D. Appleton and Co., 1870. *catalog.hathitrust.org/Record/100265213*.

Mackintosh, Fiona J. "Travellers' Tropes: Lady Florence Dixie and the Penetration of Patagonia." In *Patagonia: Myths and Realities*, edited by Fernanda Peñaloza, Jason Wilson, and Claudio Canaparo, 75–94. Oxford: Peter Lang, 2011.

Mallo, Silvia C., and Ignacio Telesa, eds., *"Negros de la patria": Afrodescendientes en las luchas por la independencia en el antiguo virreinato del Río de la Plata*. Buenos Aires: Editorial SB, 2010.

Mandrini, Raúl. "Indios y fronteras en el área pampeana (siglos XVI-XIX). Balance y perspectivas." *Anuario del IEHS* 7 (1992): 59–72.

Mandrini, Raúl, and Andrea Reguera. *Huellas en la tierra: Indios, agricultores y hacendados en la pampa bonaerense*. Tandil: IEHS, 1993.

Mansilla de García, Eduarda. *Escritos Periodísticas Completos, 1860–1892: Eduarda Mansilla de García*. Edited by Marina Guidotti. Buenos Aires: Corregidor, 2015.

______. *Lucía Miranda* (1860). Edited by María Rosa Lojo. Madrid: Iberoamericana, 2007.

______. *Recuerdos de viaje*. Edited by J. P. Spicer-Escalante. Buenos Aires: Stockcero, 2006

Mansilla, Lucio V. *Una excursión a los indios ranqueles*. Edited by J. Bouchet. Buenos Aires: Juan A. Alsina, 1890.

______. *A Visit to the Ranquel Indians*. Translated and edited by Eva Gilles. Lincoln: University of Nebraska Press, 1997.

Mármol, José. *Amalia*. Edited by Teodosio Fernández. Madrid: Cátedra, 2010.

Martínez, Alejandro. "Imágenes fotográficas sobre pueblos indígenas: Un enfoque antropológico." PhD diss., Universidad Nacional de La Plata, 2010.

Martínez Sarasola, Carlos. *Nuestros paisanos los indios: Vida, historia y destino de las comunidades indígenas en la Argentina*. Buenos Aires: Del Nuevo Extremo, 2013.

Marún, Gioconda. Introduction to *Olimpio Pitango de Monalia*, by Eduardo L. Holmberg, 7–69. Edited by Giaconda Marún. Buenos Aires: Ediciones Solar, 1994.

Mases, Enrique Hugo. *Estado y cuestión indígena: El destino final de los indios sometidos en el sur del territorio (1878–1930)*. Buenos Aires: Prometeo, 2010.

Masiello, Francine. *Between Civilization and Barbarism: Women, Nation and Literary Culture in Modern Argentina*. Lincoln: University of Nebraska Press, 1992.

Masotta, Carlos. *Indios en las primeras postales fotográficas argentinas del s. XX*. Buenos Aires: La Marca Editora, 2007.

Mataix, Remedios. "Romanticismo, feminidad e imaginarios nacionales: Las *Lucía Miranda* de Rosa Guerra y Eduarda Mansilla." *Río de La Plata* 29–30 (2006): 209–24.

Matthews, Anne. "Modern Anthropology and the Problem of the Racial Type: The Photographs of Franz Boas." *Visual Communication* 12, no. 1 (February 2013): 123–42.

Matthews, Christopher. "History to Prehistory: An Archaeology of Being Indian in New Orleans." *Archaeologies: The Journal of the World Archeological Congress* 3, no. 3 (2007): 271–95.

McClintock, Anne. *Imperial Leather: Race, Gender, and Sexuality in the Colonial Contest*. New York: Routledge, 1995.

Mills, Sara. *Discourses of Difference: An Analysis of Women's Travel Writing and Colonialism*. London: Routledge, 1993.

Mondelo, Osvaldo L. *Tehuelches: Danza con fotos*. El Calafate: El autor, 2012.

Monder, Samuel. "La ley del deseo: Acerca de 'Una excursión a los indios ranqueles,' de Lucio V. Mansilla." *Iberoamericana (nueva Época)* 8, no. 32 (December 2008): 61–74.

Montero, María. "Mansilla y sus bibliotecas." *Boletín de La Academia Argentina de Letras* 56 (1991): 103–28.

Montserrat, Marcelo, ed. *La ciencia en la Argentina entre siglos: Textos, contextos e instituciones*. Buenos Aires: Manantial, 2000.

______. "The Evolutionist Mentality in Argentina: An Ideology of Progress." In *The Reception of Darwinism in the Iberian World: Spain, Spanish America and Brazil*, edited by Thomas F. Glick, Miguel Angel Puig-Samper, and Rosaura Ruiz, 1–27. Dordrecht: Kluwer Academic Publishers, 2001.

Moraña, Mabel, Enrique Dussel, and Carlos A. Jáuregui. "Colonialism and Its Replicants." In *Coloniality at Large: Latin America and the Postcolonial Debate*, edited by Mabel Moraña, Enrique Dussel, and Carlos A. Jáuregui, 1–20. Durham: Duke University Press, 2008.

Moreno, Francisco P. "Inacayal y Foyel." *El Diario* (Buenos Aires), March 17, 1885.

______. *Reminiscencias de Francisco P. Moreno: Versión propia*. Edited by Eduardo V. Moreno. Buenos Aires: Editorial Universitaria de Buenos Aires, 1979.

______. *Viaje a la Patagonia austral, emprendido bajo los auspicios del Gobierno Nacional, 1876–1877*. Buenos Aires: Imp. de la Nación, 1879.

Musters, George Chaworth. *At Home with the Patagonians*. 2nd ed. London: John Murray, 1873. *babel.hathitrust.org/cgi/pt?id=mdp.39015018047343*.

Nagy, Mariano, and Alexis Papazian. "El campo de concentración de Martín García: Entre el control estatal dentro de la isla y las prácticas de distribución de indígenas (1871–1886)." *Corpus: Archivo virtual de la alteridad americana* 1, no. 2 (2011): 1–22. *ppct.caicyt.gov.ar/index.php/corpus/article/view/392*.

Nancuzzi, Lidia R. "'Nómades' versus 'sedentarios' en Patagonia (siglos XVIII–XIX)." *Cuadernos del Instituto Nacional de Antropología y Pensamiento Latinoamericano* 14 (1992–93): 81–92.

Nari, Marcela. *Políticas de maternidad y maternalismo político: Buenos Aires, 1890–1940*. Buenos Aires: Biblos, 2004.

Navarro Floria, Pedro. "El salvaje y su tratamiento en el discurso político argentino sobre la frontera sur, 1853–1879." *Revista de Indias* 61, no. 222 (2001): 345–76.

Navarro Floria, Pedro, Leonardo Salgado, and Pablo Azar. "La invención de los ancestros: El 'patagón antiguo' y la construcción discursiva de un pasado nacional remoto para la Argentina (1870–1915)." *Revista de Indias* 64, no. 231 (2004): 405–24.

Newton, Esther. "My Best Informant's Dress: The Erotic Equation in Fieldwork." *Cultural Anthropology* 8, no. 1 (February 1993): 3–23.

Nicoletti, María Andrea. "La congregación salesiana en la Patagonia: 'Civilizar,' educar y evangelizar a los indígenas (1880–1934)." *Estudios Interdisciplinarios de América Latina y el Caribe* 15, no. 2 (2004). *eial.tau.ac.il/index.php/eial/article/view/894*.

Nouzeilles, Gabriella. "El retorno de lo primitivo: Aventura y masculinidad." In *Entre hombres: Masculinidades del siglo XIX en América Latina*, edited by Ana Peluffo and Ignacio M. Sánchez Prado, 87–106. Madrid: Iberoamericana, 2010.

Novoa, Adriana. "The Rise and Fall of Spencer's Evolutionary Ideas in Argentina, 1870–1910." In *Global Spencerism: The Communication and Appropriation of a British Evolutionist*, edited by Bernard Lightman, 173–91. Leiden: Brill, 2015.

______. "Unclaimed Fright: Race, Masculinity, and National Identity in Argentina, 1850–1910." PhD diss., University of California, San Diego, 1988.

Novoa, Adriana, and Alex Levine. *¡Darwinistas!: The Construction of Evolutionary Thought in Nineteenth-Century Argentina*. Leiden: Brill, 2012.

______. *From Man to Ape: Darwinism in Argentina, 1870–1920*. Chicago: University of Chicago Press, 2014.

O'Connor, Erin. *Gender, Indian, Nation: The Contradictions of Making Ecuador, 1830–1925*. Tucson: University of Arizona Press, 2007.

Onelli, Clemente. *Trepando los Andes: Un naturalista en la Patagonia argentina*. Edited by Carlos Fernández Balboa. Buenos Aires: Continente, 2007.

Operé, Fernando. *Indian Captivity in Spanish America: Frontier Narratives*. Translated by Gustavo Pellón. Charlottesville: University of Virginia Press, 2008.

Pagden, Anthony. *The Fall of Natural Man: The American Indian and the Origins of Comparative Ethnology*. Cambridge: Cambridge University Press, 1982.

Palermo, Miguel Angel. "El revés de la trama: Apuntes sobre el papel económico de la mujer en las sociedades indígenas tradicionales del sur argentino." *Memoria Americana: Cuadernos de Etnohistoria* 3 (1994): 73–80.

Payró, Roberto. *La Australia argentina*. 5th ed. Buenos Aires: La Nación, 1898.

Peluffo, Ana, and Ignacio M. Sánchez Prado, eds. *Entre hombres: Masculinidades del siglo XIX en América Latina*. Madrid: Iberoamericana, 2010.

Pearce, Roy Harvey. *Savagism and Civilization: A Study of the Indian and the American Mind*. Revised edition. Berkeley: University of California Press, 1988.

Penhos, Marta. "Frente y perfil: Una indagación acerca de la fotografía en las prácticas antropológicas y criminológicas en Argentina a fines del siglo XIX y principios del XX." In *Arte y antropología en la Argentina*, edited by Marta Penhos and Marina Baron Supervielle, 17–64. Buenos Aires: Fundación Espigas, 2005.

Peñaloza, Fernanda. "On Skulls, Orgies, Virgins and the Making of Patagonia as a National Territory: Francisco Pascasio Moreno's Representations of Indigenous Tribes." BHS 87, no. 4 (2010): 455–72.

Pepe, Fernando Miguel, Miguel Añon Suarez, and Patricio Harrison. *Antropología del genocidio: Identificación y restitución. "Colecciones" de restos humanos en el Museo de La Plata.* La Plata: De la Campana, 2010.

Pepe, Fernando Miguel, Miguel Añon Suarez, Patricio Harrison, Karina Oldani, María de los Angeles Andolfo, and Marco Bufano. *"Bioiconografía": Los prisioneros de la "Campaña del desierto" en el Museo de La Plata, 1886.* La Plata: De la Campana, 2013.

Pérez Gras, María Laura. "Ojos visionarios y voces transgresoras: La cuestión del Otro en los relatos de viajes de los hermanos Mansilla." *Anales de la literatura hispanoamericana* 39 (2010): 281–304.

Perrone, Alberto. *La jirafa de Clemente Onelli.* Buenos Aires: Sudamericana, 2012.

Pesic, Peter. "Desire, Science, and Polity: Francis Bacon's Account of Eros." *Interpretation: A Journal of Political Philosophy* 26, no. 3 (Spring 1999): 333–52.

Podgorny, Irina. *El sendero del tiempo y de las causas accidentales: Los espacios de la prehistoria en la Argentina, 1850–1910.* Rosario: Prohistoria, 2009.

Podgorny, Irina, and Tatiana Kelly, eds. *Los secretos de Barba Azul: Fantasías y realidades de los archivos del Museo de La Plata.* Rosario: Prohistoria, 2012.

Podgorny, Irina, and María Margaret Lopes. *El desierto en una vitrina: Museos e historia natural en la Argentina, 1810–1890.* México: Limusa, 2008.

Potter, Elizabeth. *Feminism and Philosophy of Science: An Introduction.* London: Routledge, 2006.

Pratt, Mary Louise. *Imperial Eyes: Travel Writing and Transculturation.* New York: Routledge, 1992.

Prichard, James Cowles. *The Natural History of Man: Comprising Inquiries into the Modifying Influence of Physical and Moral Agencies on the Different Tribes of the Human Family.* London: H. Bailliere, 1843.

Quesada, Vicente G. *Memoria del ministro secretario de gobierno de la Provincia de Buenos Aires presentada a las honorables Cámaras Legislativas 1877.* Buenos Aires: Imprenta y Librerias de Mayo, 1877.

Quijada, Mónica. "Ancestros, ciudadanos, piezas de museo: Francisco P. Moreno y la articulación del indígena en la construcción nacional argentina (siglo XIX)." *Estudios interdisciplinarios de América Latina y el Caribe* 9, no. 2 (1998): 21–46.

———. "¿'Hijos de los barcos' o diversidad invisibilizada? La articulación de la población indígena en la construcción nacional argentina (siglo XIX)." *Historia mexicana* 53, no. 2 (Oct.–Dec. 2003): 469–510.

———. "La ciudadanización del 'indio bárbaro': Políticas oficiales y oficiosas hacia la población indígena de la Pampa y la Patagonia, 1870–1920." *Revista de Indias* 109, no. 217 (1999): 675–704.

———. "Los 'incas arios': Historia, lengua y raza en la construcción nacional hispanoamericana del siglo XIX." *Histórica* 20, no. 2 (1996): 243–69.

———. "Repensando la frontera sur argentina: Concepto, contenido, continuidades y discontinuidades de una realidad espacial y étnica (siglos XVIII–XIX)." *Revista de Indias* 62, no. 224 (2002): 103–42.

Qureshi, Sadiah. "Displaying Sara Baartman, the 'Hottentot Venus.'" *History of Science* xlii (2004): 233–57.

Reed-Danahay, Deborah. "Autobiography, Intimacy and Ethnography." In *Handbook of Ethnography*, edited by Paul Atkinson, Amanda Coffey, Sara Delamont, John Lofland, and Lyn Lofland, 407–25. London: Sage Publications, 2001.

Richards, Jeffrey. Introduction to *Imperialism and Juvenile Literature*. Edited by Jeffrey Richards, 1–11. Manchester: Manchester University Press, 1989.

Riegel, Henrietta. "Into the Heart of Irony: Ethnographic Exhibitions and the Politics of Difference." In *Theorizing Museums*, edited by Sharon Macdonald and Gordon Fyfe, 83–104. Oxford: Blackwell, 1996.

Roca, Julio Argentino. Archive. Correspondencia recibida, legajos 6 and 7. Archivo General de la Nación. Buenos Aires, Argentina.

Rocha, Carolina, ed. *Modern Argentine Masculinities*. Bristol: Intellect, 2013.

Rodríguez, Julia. *Civilizing Argentina: Science, Medicine, and the Modern State*. Chapel Hill: University of North Carolina Press, 2006.

Rodríguez, Mariela Eva. "De la 'extinción' a la autoafirmación: Procesos de visibilización de la comunidad tehuelche Camusu Aike (Provincia de Santa Cruz, Argentina)." PhD diss., Georgetown University, 2010. *repository.library.georgetown.edu/bitstream/handle/10822/553246/rodriguezMariela.pdf.*

Rotker, Susana. *Captive Women: Oblivion and Memory in Argentina*. Translated by Jennifer French. Minneapolis: University of Minnesota Press, 2002.

———. "*Lucía Miranda*: Negación y violencia del origen." *Revista Iberoamericana* 58, no. 178–179 (June 1997): 115–27.

Roulet, Florencia. "Mujeres, rehenes y secretarios: Mediadores indígenas en la frontera sur del Río de la Plata durante el período hispánico." *Colonial Latin American Review* 18, no. 3 (December 2009): 303–37.

Salessi, Jorge. "The Argentine Dissemination of Homosexuality, 1890–1914." *Journal of the History of Sexuality* 4, no. 3 (1994): 337–68.

Salvatore, Ricardo D. "Integral Outsiders: Afro-Argentines in the Era of Juan Manuel de Rosas and Beyond." In *Beyond Slavery: The Multilayered Legacy of Africans in Latin America and the Caribbean*, edited by Darién J. Davis, 57–80. Lanham: Rowan & Littlefield, 2007.

———. *Wandering Paysanos: State Order and Subaltern Experience in Buenos Aires During the Rosas Era*. Durham, NC: Duke University Press, 2003.

Sandweiss, Martha A. *Print the Legend: Photography and the American West*. New Haven: Yale University Press, 2002.

Sans, Mónica. "'Raza,' adscripción étnica y genética en Uruguay." *Runa* 30, no. 2 (2009): 163–74.

Sarmiento, Domingo F. *Conflicto y armonías de las razas en América*. Buenos Aires: Ostwald, 1883. *catalog.hathitrust.org/Record/000279974*.

Schávelzon, Daniel. "Argentina and Great Britain: Studying an Asymmetrical Relationship through Domestic Material Culture." *Historical Archaeology* 47, no. 1 (2013): 10–25. *www.jstor.org/stable/43491296*.

Schoolcraft, Henry Rowe. *Algic Researches: North American Indian Folktales and Legends*. Edited by Curtis M. Hinsley. Mineola, NY: Dover Publications, 1999.

Secord, James. "Knowledge in Transit." *Isis* 95, no. 4 (December 2004): 654–72.

Seluja Cecín, Antonio, and Alberto Paganini. *"Tabaré": Proceso de creación*. Montevideo: Biblioteca Nacional, 1979.

Sera-Shriar, Efram. "Anthropometric Portraiture and Victorian Anthropology: Situating Francis Galton's Photographic Work in the Late 1870s." *History of Science* 53, no. 2 (2015): 155–79.

Shkatulo, Olena. "Nation, Gender, and Beyond in Mansilla's *Recuerdos de Viaje*." *Journal of Iberian and Latin American Research* 23, no. 1 (2017): 62–74. DOI: 10.1080/13260219.2017.1308000.

Siegrist de Gentile, Nora, and María Haydée Martín. *Geopolítica, ciencia y técnica a través de la campaña del desierto*. Buenos Aires: Editorial Universitaria de Buenos Aires, 1981.

Sivasundaram, Sujit. "Sciences and the Global: On Methods, Questions, and Theory." *Isis* 101, no. 1 (March 2010): 146–58.

Smith, Michelle. *Empire in British Girls' Literature and Culture: Imperial Girls, 1880–1915*. New York: Palgrave/Macmillan, 2011.

Socolow, Susan Migden. "Spanish Captives in Indian Societies: Cultural Contact along the Argentine Frontier, 1600–1835." *Hispanic American Historical Review* 72, no. 1 (1992): 73–99.

Soler, Mariano. *La masonería y el catolicismo: Estudio comparado bajo el aspecto del derecho común, las instituciones democráticas y filantrópicas, la civilización y su influencia social*. Montevideo: Andrés Ríus, 1885. *ia802305.us.archive.org/27/items/catolicosymasoneoosole/catolicosymasoneoosole.pdf*.

Sollors, Werner. *Neither Black nor White yet Both: Thematic Explorations of Interracial Literature*. Oxford: Oxford University Press, 1997.

Somerville, Siobhan. "Scientific Racism and the Emergence of the Homosexual Body." *Journal of the History of Sexuality* 5, no. 2 (October 1994): 243–66.

Sommer, Doris. *Foundational Fictions: The National Romances of Latin America*. Berkeley: University of California Press, 1991.

Spillane McKenna, Maureen. "El debate del ser nacional argentino: La frontera en las obras de Lucio Mansilla y César Aira." *Hispanófila* 142 (September 2004): 89–100.

Stein, Melissa N. *Measuring Manhood: Race and the Science of Masculinity, 1830–1934*. Minneapolis: University of Minnesota Press, 2015.

Stepan, Nancy. "Biological Degeneration: Races and Proper Places." In *Degeneration: The Dark Side of Progress*, edited by J. Edward Chamberlin and Sander L. Gilman, 97–120. New York: Columba University Press, 1985.

______. *"The Hour of Eugenics": Race, Gender, and Nation in Latin America*. Ithaca, NY: Cornell University Press, 1991.

Stocking, George W., Jr. *Race, Culture, and Evolution. Essays in the History of Anthropology*. New York: Free Press, 1968.

Stoler, Ann Laura. *Carnal Knowledge and Imperial Power: Race and the Intimate in Colonial Rule*. Berkeley: University of California Press, 2002.

Syed, Jawad, and Faiza Ali. "The White Woman's Burden: From Colonial *Civilisation* to Third World *Development*." *Third World Quarterly* 32, no. 2 (March 2011): 349–65. DOI: 10.1080/01436597.2011.560473.

Szurmuk, Mónica. *Women in Argentina: Early Travel Narratives*. Gainesville: University Press of Florida, 2000.

Tamagnini, Marcela. *Soberanía: Territorialidad indígena. Cartas civiles II*. Temuco: Centro de Documentación Mapuche, 2003. *www.mapuche.info/wps_pdf/tamagnini031104.pdf*.

ten Kate, Herman. "Matériaux pour servir à l'anthropologie des Indiens de la République Argentine." *Revista del Museo de La Plata* 12 (1906): 33–57.

Thompson, Elizabeth. *The Pioneer Woman: A Canadian Character Type*. Montreal: McGill-Queen's University Press, 1991.

Topcic', D. Osvaldo. *Historia de la Provincia de Santa Cruz: Crónicas y testimonios.* Córdoba: Centro de Estudios Históricos "Prof. Carlos S. A. Segreti," 2006.

Torre, Claudia. *Literatura en tránsito: La narrativa expedicionaria de la Conquista del Desierto.* Buenos Aires: Prometeo, 2010.

Trelles, Manuel Ricardo. "Al señor ministro de gobierno Doctor Don Nicolás Avellaneda." *La revista de Buenos Aires* 16 (1868): 589–92.

Unzueta, Fernando. "Scenes of Reading: Imagining Nations / Romancing History in Spanish America." In *Beyond Imagined Communities: Reading and Writing the Nation in Nineteenth-Century Latin America*, edited by Sara Castro-Klarén and John Charles Chasteen, 115–60. Washington, DC: Woodrow Wilson Center Press, 2003.

Valko, Marcelo. *Cazadores de poder: Apropiadores de indios y tierras (1880–1890).* Buenos Aires: Ediciones Continente, 2015.

______. *Pedagogía de la desmemoria: Crónicas y estrategias del genocidio invisible.* Buenos Aires: Continente, 2013.

Videla, Liliana. "María, la cacica de los tehuelches." *Todo es historia* 477 (2007): 28–35.

Vignati, Milcíades Alejo. "Falacias iconográficas." *Relaciones de la Sociedad Argentina de Antropología* 4 (1944): 259–62.

______. "Iconografía aborigen I: Los caciques Sayeweke, Inakayal y Foyel y sus allegados." *Revista del Museo de La Plata (nueva Serie)* 2, Antropología, no. 10 (1942): 13–48.

Viñas, David. *Indios, ejército y frontera.* Mexico: Siglo Veintiuno, 1982.

Vitulli, Juan M., and David M. Solodkow. "Ritmos diversos y secuencias plurales: Hacia una periodización del concepto 'criollo.'" In *Poéticas de lo criollo: La transformación del concepto "criollo" en las letras hispanoamericanas (siglo XVI al XIX)*, 9–58. Edited by Juan M. Vitulli and David M. Solodkow. Buenos Aires: Corregidor, 2009.

Wade, Peter. *Race and Sex in Latin America.* London: Pluto Press, 2009.

Young, Robert. *Colonial Desire: Hybridity in Theory, Culture and Race.* London: Routledge, 1995.

Zeballos, Estanislao S. *Callvucurá y la dinastía de los Piedra.* 3rd ed. Buenos Aires: Jacobo Peuser, 1890.

______. *La conquista de quince mil leguas: Ensayo para la ocupación definitiva de la Patagonia (1878).* Buenos Aires: Ediciones Continente, 2008.

______. *Descripción amena de la República Argentina. Tomo 1. / Viaje al país de los Araucanos.* Buenos Aires: Jacobo Peúser, 1881.

______. *Painé y la dinastía de los Zorros.* 2nd ed. Buenos Aires: Jacobo Peuser, 1889. *books.google.com/books/?id=-OE9AQAAIAAJ.*

______. *Relmú, reina de los pinares.* 2nd ed. Buenos Aires: Jacobo Peuser, 1894. *books.google.com/books?id=UD9UAAAAMAAJ&hl=en.*

Zorrilla de San Martín, Juan. "Darwin." In *Juan Zorrilla de San Martín en la prensa: Escritos y discursos*, by Juan Zorrilla de San Martín, edited by Antonio Seluja Cecín, 47–50. Montevideo: Comisión Nacional de Homenaje del Sesquicentenario de los Hechos Históricos de 1825, 1975.

______. "Descubrimiento y conquista del Río de la Plata." In *Juan Zorrilla de San Martín: Conferencias y discursos*, vol. 1, edited by Benjamín Fernández y Medina, 49–99. Montevideo: Imprenta Nacional Colorada, 1930.

______. "A mi América española." In *Huerto Cerrado*, 313–27. Montevideo: Imprenta Nacional Colorada, 1930.

______. *Tabaré*. Edited by Antonio Seluja Cecín. Montevideo: Universidad de la República, 1984.

Zorrilla de San Martín, Juan, and Miguel de Unamuno. *Correspondencia de Zorrilla de San Martín y Unamuno*. Edited by Arturo Sergio Visca. Montevideo: Instituto Nacional de Investigaciones y Archivos Literarios, 1955.

INDEX

Page numbers in **bold** indicate figures.